SUSTAINABLE
FUTURES

Dedication

In memory of Frank J Fenner

The world is today confronted by many intractable and interconnected problems: a rapidly rising human population and expectations for a better life, diminishing energy resources with which to meet increased food production and the prospect of rising temperatures and their consequences.

The 2013 Fenner Conference on the Environment, held at the Australian Academy of Science on 10–11 October 2013, brought together a panel of expert speakers to address these urgent matters under the title *Population, Resources and Climate Change: Implications for Australia's Near Future.*

Organised by Sustainable Population Australia (SPA) and opened by the President of the Australian Academy of Science, Professor Suzanne Cory AC, the Conference was then addressed by the keynote speaker, Professor Paul Ehrlich, a pioneer in alerting the world to the problems of overpopulation. Ehrlich rose to prominence with his 1968 book *The Population Bomb.*

The 20 speakers that followed each addressed one of the three topics from the perspective of their own expertise and also made connections with at least one of the other topics. These connections were easy to make in many cases, for instance, those between resources and climate change. As conventional oil supplies decline and are replaced by oil from unconventional sources, more energy is required to produce the same amount of oil. Generally more greenhouse gases are emitted in the process, worsening climate change.

At the end of the conference there was a strong feeling that the content of the excellent papers should reach a wider audience. CSIRO Publishing agreed to publish a book on the topic that would reflect the content of the papers and the discussion that followed each session. All but two of the speakers were able to provide a version ready for publication, which it is hoped will help in the development of public policy to address these immensely urgent matters.

Jenny Goldie
National President, Sustainable Population Australia
Co-editor

SUSTAINABLE
FUTURES

Linking population, resources and the environment

EDITORS: JENNY GOLDIE AND KATHARINE BETTS

CSIRO

PUBLISHING

National Library of Australia Cataloguing-in-Publication entry

Goldie, Jenny, editor.

Sustainable futures: linking population, resources and the environment/Jenny Goldie, Katharine Betts

9781486301898 (paperback)
9781486301904 (epdf)
9781486301911 (epub)

Sustainable development – Australia.
Natural resources – Australia.
Environmental policy – Australia.
Economic development – Environmental aspects – Australia.
Australia – Population – Environmental aspects.

Betts, Katharine, editor.

338.927

Published by

CSIRO Publishing
150 Oxford Street (PO Box 1139)
Collingwood VIC 3066
Australia

Telephone: +61 3 9662 7666
Local call: 1300 788 000 (Australia only)
Fax: +61 3 9662 7555
Email: publishing.sales@csiro.au
Web site: www.publish.csiro.au

Front cover (left to right): jokerpro (Shutterstock.com); kaband (Shutterstock.com); B747 (Shutterstock.com)

Set in 11/13.5 Adobe Minion Pro and Helvetica Neue LT Std
Edited by Anne Findlay, Editing Works Pty Ltd
Cover design by James Kelly
Typeset by Desktop Concepts Pty Ltd, Melbourne
Index by Bruce Gillespie
Printed in China by 1010 Printing International Ltd

Original print edition:
The paper this book is printed on is in accordance with the rules of the Forest Stewardship Council®. The FSC® promotes environmentally responsible, socially beneficial and economically viable management of the world's forests.

MIX
Paper from responsible sources
FSC® C016973

Contents

Author biographies

Dr Katharine Betts is Adjunct Associate Professor of Sociology at Swinburne University of Technology and was Joint-Editor of the demographic journal *People and Place* from 1993 to 2010. She wrote *Ideology and Immigration* (MUP 1988) published in a second edition as *The Great Divide* (Duffy and Snellgrove 1999). Recent works include articles on attitudes to immigration and population growth in Australia (2010) and the demographic and social implications of ageing (her most recent paper is entitled 'The ageing of the Australian population: triumph or disaster', an online publication through the Monash Centre for Population and Urban Research, April 2014). She was a member of the Sustainable Population Sustainable Development Panel chaired by Bob Carr, in 2010, and served on the National Council of the Australian Population Association from 1996 to 2000, and from 2002 to 2006.

Dr Bob Birrell is Reader in Sociology at Monash University. He was the founding director of the Centre for Population and Urban Research (CPUR) at Monash University and founder and Joint-Editor of the demographic journal *People and Place* from 1993 to 2010. He has advised successive Australian governments on immigration policy, most recently as part of the Evaluation of the General Skilled Migration Categories which reported in 2006. His work covers the environmental, social and economic implications of population growth in Australia. Recent publications include *Immigration and the Resources Boom Mark 2* (CPUR 2011), *The End of Affordable Housing in Melbourne?* (CPUR 2012) and *Immigration Overshoot* (CPUR 2012).

Dr Paul Collins is a historian, broadcaster and writer. A Catholic priest for 33 years, he resigned from the active priestly ministry in 2001 due to a dispute with the Vatican over his book *Papal Power* (1997). He is the author of 13 books, the most recent of which, *The Birth of the West,* was published in New York in February 2013. He is well known as a commentator on Catholicism and the papacy and he also has a strong interest in ethics, environmental and population issues. He has a Master's degree in theology (ThM) from Harvard University, and a PhD in history from the Australian National University. He lives in Canberra.

Mr Julian Cribb is an author, journalist, editor and science communicator. He is principal of Julian Cribb & Associates which provides specialist consultancy in the communication of science, agriculture, food, mining, energy and the environment. His career includes appointments as newspaper editor, scientific editor for *The Australian* newspaper, director of national awareness for CSIRO, member of numerous scientific boards and advisory panels, and president of national professional bodies for agricultural journalism and science communication. His published work includes over 8000 articles, 3000 media releases and eight books. He has received 32 awards for journalism. His internationally acclaimed book *The Coming Famine* explores the question of whether we can feed humanity through the mid-century peak in numbers and food demand. His latest book, *Poisoned Planet* (A&U 2014), explores the impact of 150 billion tonnes of human chemical emissions on the human race and all life on Earth.

Professor Chris Dickman has long been fascinated by patterns in biological diversity and in the factors that affect it. His current work focuses on biota in arid environments and on a range of other projects in applied conservation and management. An ARC Professorial Fellow, Professor Dickman has been a prolific trainer of postgraduates, supervising 42 Honours, 37 Masters and 52 PhD students over the last 25 years. He has written or edited 20 books and monographs and authored a further 330 journal articles and book chapters. He is the recipient of several national and international awards, and was the inaugural chair of the NSW Government Scientific Committee from 1996 to 2002.

Dr Rhondda Dickson has been Chief Executive of the Murray–Darling Basin Authority (MDBA) since June 2011. She has over 20 years' experience working with states and territories in the development and implementation of natural resource management policies. Since joining MDBA, Dr Dickson has led the development of a significant piece of national water reform – the Basin Plan. Her focus has been on striking a genuine balance between the environmental, social and economic needs of the Basin and its people. Prior to joining MDBA, she was closely involved in the development of the National Action Plan for Water Quality and Salinity, national forest policy and national approaches to vegetation management.

Mr Ian Dunlop is a Cambridge-educated engineer with a particular interest in the interaction of corporate governance, corporate responsibility and sustainability. He is a former international oil, gas and coal industry executive. He chaired the Australian Coal Association in 1987–88, chaired the Australian Greenhouse Office Experts Group on Emissions Trading from 1998 to 2000 and was CEO of the

Australian Institute of Company Directors from 1997 to 2001. Mr Dunlop is a Director of Australia 21, Chairman of Safe Climate Australia, a Member of the Club of Rome and Fellow of the Centre for Policy Development. He advises internationally on climate, energy and sustainability.

Anne H. Ehrlich is a senior research scientist emeritus and former associate director for policy of the Center for Conservation Biology, in the Biology Department at Stanford University. She has carried out research and coauthored many technical articles in population biology and ecology and has taught seminar courses on environmental policy at Stanford since 1981. She has written extensively on issues of public concern such as population, environmental protection and the environmental consequences of nuclear war, and has coauthored more than a dozen books. Anne is a fellow of the American Academy of Arts and Sciences, and her honours include two honorary doctorate degrees, the United Nations Environmental Programme/Sasakawa Prize, the Heinz Award for Environmental Achievement, and the Tyler Prize (all shared with Paul Ehrlich).

Professor Paul Ehrlich is Bing Professor of Population Studies and President of the Center for Conservation Biology at Stanford University, Stanford, California. He was a pioneer in alerting the public to the problems of overpopulation, and in raising issues of population, resources and the environment as matters of public policy. He has published 43 books, 13 of which are co-authored with Anne Ehrlich, and more than 600 scientific papers. He rose to international prominence with his 1968 book, *The Population Bomb*. He has received several honorary degrees and is the recipient of numerous awards including in 2001 the Eminent Ecologist Award of the Ecological Society of America and the Distinguished Scientist Award of the American Institute of Biological Sciences.

Ms Jenny Goldie is a former science teacher and science communicator. She was a founding member of Australians for an Ecologically Sustainable Population (now Sustainable Population Australia) and is currently SPA's national president. She was sufficiently influenced by Paul Ehrlich's *The Population Bomb* to limit her biological children to one and to adopt or foster the rest. She is also active in climate change and peak oil groups. She was co-organiser of the 2013 Fenner Conference and co-editor of the book *In Search of Sustainability* (2005) that arose out of a conference by the same name.

Major General the Honourable Michael Jeffery, AC, AO(Mil), CVO, MC (Retd) graduated from the Royal Military College, Duntroon, into Infantry, serving operationally in Malaya, Borneo, Papua New Guinea and Vietnam, where he was awarded the Military Cross and the South Vietnamese Cross of Gallantry. From 1993 to 2000 he was the Governor of Western Australia. From 2003 to 2008 he served as Governor-General of Australia where his key interests were in youth, education and landscape regeneration. He is Chairman of Future Directions International (which he founded), Outcomes Australia, and Soils for Life and is patron of numerous charitable organisations. He was recently appointed by the Prime Minister as the National Advocate for Soil Health and as the Australian Envoy for The Queen Elizabeth Diamond Jubilee Trust.

Professor Gary Jones is Chief Executive of eWater, a not for profit river basin management and modelling organisation (formerly eWater CRC), and an Adjunct Professor with the University of Canberra, Institute of Applied Ecology. He is also Chairman of the International RiverFoundation, a charitable organisation dedicated to the restoration and protection of the world's rivers. Previously he was a Senior Principal Research Scientist with CSIRO Australia, a Senior Research Fellow at the University of Newcastle Upon Tyne, UK and a Fulbright Postdoctoral Fellow at MIT in Boston USA. He is the author of over 100 scientific publications on topics including the chemistry and ecology of toxic blue-green algae, and environmental flows science and management.

Dr Michael Lardelli received his PhD in Developmental Genetics from the UK Council for National Academic Awards in 1991. After six years of postdoctoral work in Sweden he returned to Australia in 1997 and is currently Senior Lecturer in Genetics at the University of Adelaide where his research focuses on understanding the molecular basis of Alzheimer's disease. He has been active in spreading awareness of peak oil since 2004 and worked with Professor Kjell Aleklett to produce the English version of his book *Peeking at Peak Oil*, published in 2012. He is currently a member of the executive committee of SPA.

Professor David Lindenmayer AO is Professor of Ecology at the Fenner School of Environment and Society at The Australian National University. He is a leading conservation biologist, contributing significantly to the understanding of biodiversity, both within Australia and around the world. He specialises in establishing large-scale, long-term research programs that are underpinned by rigorous experimental design, detailed sampling and innovative statistical analyses. Professor Lindenmayer has published 36 books and over 900 scientific articles on wildlife ecology, forest ecology and management, woodland ecology and conservation biology. He has worked on biodiversity conservation for more than 30 years. He was elected to the Australian Academy of Science in 2008 and has won numerous environmental and conservation awards.

Professor Ian Lowe AO is an emeritus professor in the School of Natural Sciences at Griffith University and was president of the Australian Conservation

Foundation from 2004 to 2014. His principal research interests are in policy decisions influencing the use of energy, science and technology; energy use in industrialised countries; large-scale environmental issues and sustainable development. In 1988 he was director of the Commission for the Future. He chaired an advisory council that produced the first national state of the environment report in 1996, and delivered the ABC Boyer Lectures in 1991. In 2002 Professor Lowe was awarded a Centenary Medal for contributions to environmental science and won the Eureka Prize for the promotion of science.

Professor Anthony McMichael AO is Emeritus Professor (Population Health) at ANU. His primary research and policy interests relate to social and environmental influences on health. His work has spanned diet/nutrition and disease, urban air pollution, and environmental lead impacts on child development. Since 1990 he has focused on the population health impacts of human-induced environmental changes, especially climate change. He is an elected member of the US National Academies of Science; has co-chaired scientific assessments by the Intergovernmental Panel on Climate Change (IPCC); and chairs a WHO expert group on joint influences of environment, climate and agriculture in infectious disease emergence. He has published widely on these topics.

Dr Simon Michaux has a BAppSc in Physics and Geology and a PhD in mining engineering. He has worked in the mining industry for 18 years in various capacities. He has worked in industry-funded mining research, coal exploration and in the commercial sector in an engineering company as a consultant. Areas of technical interest have been: geometallurgy; mineral processing in comminution, flotation and leaching; blasting; mining geology; geophysics; feasibility studies; mining investment; and industrial sustainability.

Ms Sharyn Munro is a literary activist, reaching beyond the converted with her personal form of nature and environmental writing. Essayist, award-winning short story writer and author of three non-fiction books, she has deep sympathy for the people dispossessed and the places destroyed by the resources rush. Ms Munro lives in a solar-powered mudbrick cabin on her Upper Hunter wildlife refuge, where her books *The Woman on the Mountain* and *Mountain Tails* are set. Concern for the future of her grandchildren drove her to research and write the very different *Rich Land, Wasteland – How Coal is Killing Australia*, covering coal and CSG (coal seam gas).

Mr Mark O'Connor is a poet and environmentalist. He is an editor of the Oxford University Press textbook *Protected Area Management*, author of *This Tired Brown Land*, co-author of *Overloading Australia* and of *Big Australia? Yes/No* (Pantera

Press 2012). He has published more than a dozen books of verse and received several awards. With Judith Wright, he founded Writers for an Ecologically Sustainable Population. He has taught at James Cook University, Aarhus University, and the Australian National University (ANU), and has been the ANU's HC Coombs Fellow, and the Museum of Victoria's Thomas Ramsay Science and Humanities Fellow.

Dr Jane O'Sullivan has had a career as an agricultural scientist, with a focus on soil fertility management in the South Pacific and South-East Asia. She has long been aware of the impacts of population pressure on land resources, and the need to match efforts to increase food production with efforts to limit demand. Realising that population policy is based on economic claims rather than ecology, she has been researching and writing about population economics since 2008. Her research has challenged widely attested beliefs about the 'demographic transition', the impact of ageing on workforce and the '3Ps' of population, participation and productivity. She is most widely recognised for quantifying the infrastructure cost of population growth.

Professor Roger Short held a number of academic posts in the UK and US before taking up a Personal Chair in Reproductive Biology at Monash University in 1982. From 1996 to 2005 he was Wexler Professorial Fellow at the Royal Women's Hospital. Since then he has been an Honorary Professorial Fellow in the Faculty of Medicine, Dentistry and Health Sciences at the University of Melbourne, where he is still actively involved in research and teaching reproduction to science and medical students. He has published over 300 scientific papers and his most recent book, co-authored with Dr Malcolm Potts, was *Ever Since Adam and Eve: the Evolution of Human Sexuality*.

Mr Kelvin Thomson MP, a former Member of the Victorian Parliament, has been Federal Member for Wills since 1996. He held a number of Shadow portfolios during the Howard years, including being Shadow Environment Minister between 2001 and 2004 when he was responsible for Labor's adoption of policies to tackle climate change. These included ratification of the Kyoto Protocol, introduction of an emissions trading system and lifting the Renewable Energy Target. In 2009, he kick-started a national debate concerning Australia's population with a speech to Parliament and the release of a 14 Point Plan for Population Reform. He was appointed as the Parliamentary Secretary for Trade in February 2013.

Dr Haydn Washington is a Visiting Fellow at the Institute of Environmental Studies, UNSW. He is an environmental scientist with a 35-year history in environmental science. He has worked extensively within environmental non-government organisations, including being a councillor on the ACF for four terms and secretary of the Colo Committee, which led the campaign to create Wollemi National Park. As an environmental writer, he has written five books on the environment: *Ecosolutions: Environmental Solutions for the World and Australia* (1991), *A Sense of Wonder* (2002), *The Wilderness Knot* (2009), *Climate Change Denial: Heads in the Sand* (2011) and *Human Dependence on Nature* (2013). He is keenly interested in *why* societies deny environmental problems.

Introduction

Jenny Goldie

In late 2013 a report was published that should have had alarm bells ringing loud and hard. A study by researchers at the University of Nebraska found that around 30 per cent of the main global cereal crops, including corn, rice and wheat, displayed an abrupt decrease in yields or had plateaued despite an increase in investment in agricultural research and development, education and infrastructure (Grassini *et al.* 2013). The study suggested that maximum potential yields under the industrial model of agribusiness had already occurred.

This would be worrying enough had global population numbers levelled off, but they had not. Global population at the end of 2013 was 7.2 billion and growing at a rate of 1.14 per cent. That translates to 82 million more people every year and, according to the revised projections from the United Nations, world population will be 9.6 billion by 2050 (United Nations, Department of Economic and Social Affairs, Population Division 2013).

Indeed, two years before, the United Nations' Food and Agricultural Organization (FAO) had found that farmers would have to produce 70 per cent more food by 2050 to feed the anticipated world population (FAO 2011). Yet, the FAO report acknowledged that a quarter of farmland is already highly degraded and warned that the trend needed to be reversed. As most available farmland is already being farmed, a major 'sustainable intensification' of agricultural productivity on existing farmland would be necessary. The report noted that climate change coupled with poor farming practices was leading to a loss of productivity.

The University of Nebraska study implies that, if there is little new land to farm and yields of major crops in many areas are plateauing or declining under the industrial model of farming, achieving the required 70 per cent increase in food production by 2050 will be very difficult indeed.

As 2014 broke, a paper from the University of NSW, published in *Nature*, predicted temperatures are on course to rise at least 4°C by the end of the century with potentially catastrophic results for agriculture (Sherwood *et al.* 2014). Earlier climate models had assumed clouds would limit temperature increases and projected smaller rises but the study found, in fact, clouds had limited effect in cooling. This came as Australia ended its hottest year in a century of temperature records (Bureau of Meteorology 2014).

These two papers and the FAO report illustrate that the world is facing some intractable problems that are often interconnected. We are approaching limits to food production even as the global population grows inexorably, land and environment degrade and temperatures rise. We are facing limits to growth on a number of fronts.

This book is about population, resources and environment and the links between them. Its authors are the speakers from the 2013 Fenner Conference on Environment who addressed the implications for Australia's near future of population size and growth, resources – both mineral and natural – and climate change. The title of the book has been broadened to 'environment' from 'climate change', because the conference addressed many other environmental issues such as biodiversity and not just climate change.

When Sustainable Population Australia (SPA),[1] the organisation that ran the Fenner Conference, was formed in 1988, its prime aim was to make the public aware of the limits of Australia's population growth from an environmental viewpoint. It was evident that Australia's population, then over 16 million, was damaging the environment, particularly as habitat was destroyed to make way for urban and agricultural expansion. Renowned population expert Paul Ehrlich and Tim Flannery, later Australian of the Year and SPA Patron, suggested at the time that Australia had already exceeded its carrying capacity.

Soon after SPA's formation, however, it was evident that human carrying capacity was going to be affected by another phenomenon – climate change. The now late Stephen H Schneider, adviser to seven US presidents on climate change, spoke in Parliament House, Canberra, stressing the need to reduce greenhouse gas emissions as a means of combating global warming. While we worried then, it was worry for future generations. We could not imagine that climate change would, a quarter century later, be an existential problem affecting us here on the planet right now. With 0.8°C warming since pre-industrial times, the bell-curve had shifted sufficiently to the right for extreme weather events to become more frequent and more extreme in intensity, affecting millions of people.

In 1998, yet another issue emerged that would seriously affect not only carrying capacities, but the world's economies. An article in *Scientific American* by Colin Campbell and Jean Laherrère, 'The end of cheap oil', alerted us to the looming issue of peak oil (Campbell and Laherrère 1998). In the few years following, a whole raft of books including *The Party's Over* (Heinberg 2003), *The Long Emergency* (Kunstler 2005) and *Beyond Oil* (Deffeyes 2005) largely confirmed what Campbell and Laherrère had postulated, namely, that global oil production was soon to peak. According to the International Energy Agency, conventional oil peaked in 2006 (IEA 2010). The emergence of unconventional oil is extending the total oil peak possibly by a decade; nevertheless, the problem has not gone away.

It was entirely appropriate that Paul Ehrlich, who had raised the issues of population, resources and the environment as matters of public policy for the past several decades, should deliver the keynote address at the 2013 Fenner Conference. He argued that overpopulation, along with overconsumption by the rich, is the critical root cause of the human predicament. Ehrlich expands on this in Chapter 1, co-authored with his wife Anne.

The next session addressed the environmental implications of population growth. How fortunate we were to have had three of Australia's leading biologists – David Lindenmayer, Hugh Possingham and Chris Dickman – articulate so clearly what SPA had tried to get across for 25 years.

Lindenmayer noted, as he does in Chapter 2, that Australia and the rest of the world face enormous environmental challenges. A rapidly expanding population demands access to natural resources but these are increasingly difficult to find, be they food and fibre from degraded forests and farms, or oil from increasing depths below the ocean. Human demands have huge negative impacts on biodiversity, not only at the margins of our cities where new suburbs encroach on endangered woodland, but in more remote places where resource extraction and use affect environments and the biodiversity that occurs within them.

Illustrating this point, Sharyn Munro, later in the day and again in this book in Chapter 9, described in graphic detail the ravages wrought on the countryside in New South Wales and Queensland by the coal and the coal seam gas industries. Munro warns of the 'environmental atrocity if Clive Palmer is allowed to mine the Bimblebox Nature Refuge there and wipe out its richly diverse flora and fauna'. Yet on the Friday before Christmas 2013, Greg Hunt, the Federal Minister for the Environment, gave his conditional approval for Palmer's Galilee Project and, in turn, the green light for the destruction of Bimblebox Nature Refuge.

Possingham reinforced Lindenmayer's message that the environmental impacts of population growth on biodiversity are vast. On the other hand, we know very little about many of the potentially positive environmental impacts of our interventions, such as dedicating national parks, and how they might work. To overcome this lack of knowledge, Possingham and other conservation researchers are involved in the Environmental Decisions Group (EDG) (<http://www.edg.org. au/>) and work on the science of effective decision-making to better conserve biodiversity. Regrettably, Possingham was unable to provide a chapter for this book but his and others' work can be found at the EDG website and through their journal *Decision Point* (<www.decision-point.com.au>).

Dickman addressed, and again in Chapter 3, the relationship between human population growth and wildlife in Australia. Wildlife has indeed fared poorly in the presence of expanding populations, be they pre-1788 or post-1788. Dickman showed there is a tight correlation between past terrestrial vertebrate extinctions and number of people in Australia, and from this we can extrapolate the number of new extinctions that are likely to occur should population grow according to official projections – these may be as high as 70 million by 2101 (ABS 2013). These figures on extinctions are disturbing enough but perhaps even worse is the problem of cultural memory loss whereby our connection to the continent's natural environment collectively diminishes as we become more urbanised. If there is little or no appreciation of the continent's natural riches, we must anticipate accelerating loss of wildlife and other species 'as we look the other way'.

Although SPA is intrinsically environmental in its focus, it also seeks to make the public aware of the limits of Australia's population growth from social and economic viewpoints. In the session devoted to this, Bob Birrell said that with net migration currently responsible for 60 per cent of Australia's annual population growth and successive governments delegating immigration selection to employers, achieving a sustainable population policy is difficult (Chapter 4). Mark O'Connor (Chapter 5) argued that stabilising population would assist us to relieve overstretched infrastructure, ease cost of living pressures, promote education, training and employment of young adults, minimise high rise and sprawl, and create a more resilient economy that does not depend on the resources boom. Jane O'Sullivan (Chapter 6) argued that 'ageing paranoia', that is, the fear of demographic ageing, is unwarranted since in the two 'oldest' countries, Germany and Japan, there has been greater workforce participation and, in turn, lower levels of income inequality. These are 'depopulation dividends'.

As well as Sharyn Munro in the session on *Resources – Oil, Minerals, Coal and Gas*, there were two other remarkable presentations. Michael Lardelli (Chapter 7) spoke about peak oil (the peak rate of oil production, not resource size) and Australia's growing vulnerability to oil shocks. Lardelli referred to the infamous BITRE 117 report of 2009 warning of such vulnerability that was suppressed by the Australian government. Simon Michaux (Chapter 8) addressed the challenges facing the mining industry. The biggest one is that, in the past 10 years, 40 per cent more energy, capital, labour etc. is required to get one unit of metal out of the ground. As the costs of energy and water increase in coming years, the very viability of mining will be affected, leading to a decline in mining that may well become permanent.

Oil, minerals, coal and gas are non-renewable resources (NRRs) but soil and water are renewable and require a different approach. Three leaders in their respective fields addressed *Food, Land and Water – a Blueprint for the Future*. Michael Jeffery (Chapter 10) felt that feeding the nine billion or more by 2050 presents one of the great challenges of our time because of decreasing availability of agricultural land, rapidly diminishing aquifer sourced water, degraded landscapes and increasing costs of fossil fuel-based fertilisers. To save the planet, he said we must save the soil 'with alacrity and focus'. Rhondda Dickson (Chapter 11) described how the Murray–Darling Basin Plan established an environmentally sustainable level of take that, along with market reforms, allows for an adaptable and flexible approach to dealing with the consequences of climate change as they emerge. Gary Jones (Chapter 12) looked at the concept of developing the north as 'The Food Bowl of Asia'. He said we could increase production there if we learned our lessons from agriculture in the south, but we need to do it selectively, wisely and with broad environmental and cultural stewardship.

The session on climate change at the conference was confronting. Michael Raupach, who also regrettably, was unable to contribute to this book, showed the now famous saw-tooth graph of carbon dioxide, methane and sea levels (retrieved from ice-core records) rising and falling in synchrony over the last 800 000 years, and also temperature (inferred from the isotopic composition of both the ice and the air in ice bubbles). Through these Ice Age cycles, oscillations of the order of five degrees occurred but even at the highest point, carbon dioxide never reached 300 parts per million (ppm). We are now, however, at about 400 ppm, relative to a pre-industrial background of 280 ppm. And carbon dioxide is going up at about two parts per million per year and showing no signs of coming off that growth rate. Looking at a number of climate indicators since 1850, ocean air temperature shows a rising trend, as does sea level, which rose by about 20 cm over the last century and is still rising by about 3 mm per year. An important climate indicator is the extent of Arctic sea ice, which a couple of years ago reached a record minimum that surprised everybody. It did not reach as low a level in 2013, but the continuing trend is absolutely clear: the sea ice is contracting, and that has massive consequences for both ecosystems and climate in the Arctic. Raupach argued that if we are to avert dangerous climate change, a cap must be put on emissions, that is, a cap on the amount of fossil fuel that can be burned. To give us a one-in-two chance of success of staying below two degrees of warming above pre-industrial levels, the allowed budget of cumulative CO_2 emissions from 1750 to the far future is about 1000 billion tonnes (bt) of carbon. If we want a two-in-three chance, then our budget goes down to 800 bt. We have used 550 bt at the moment and we are using about 10 bt per year, so we have about 25 years to go before we hit our carbon budget – before we have exhausted our quota. (As Ian Dunlop notes in Chapter 14, the Climate Commission had given an even tighter time frame of 15 years, based on the world having a 75 per cent chance of staying within the 2 degree guardrail. Not factored into any of these calculations, however, are possible climate feedbacks such as release of methane from the Arctic permafrost.) The question is: how do we share this carbon budget between nations? However we do it, it must be done with the twin goals of environmental sustainability and social equity in mind.

Also in this session, Anthony McMichael (Chapter 13) warned that unabated climate change threatens the ecological and social foundations of population health and survival by affecting infectious disease patterns, food yields, freshwater supplies and more. Ian Dunlop (Chapter 14), former oil, gas and industry executive, said we are on a trajectory for 4°C warming yet warned that business in a 4°C world is not possible. While the technologies involved in developing unconventional oil supplies are impressive, they all suffer from the law of diminishing returns as their energy returned on energy invested (EROEI) drops.

There are, of course, many obstacles to action in dealing with these emerging problems. Religion, denialism, politicians and the media are all culpable. At the same time, there are opportunities for action. Paul Collins, Haydn Washington, Kelvin Thomson and Julian Cribb addressed these obstacles and opportunities in Chapters 15–18.

In Chapter 19, Professor Roger Short provides his views on how we might achieve an end to population growth, an absolute necessity if we are to have a sustainable future. In the concluding chapter, SPA Patron Professor Ian Lowe, who had been such a supportive presence throughout, reviews the conference and includes the Conference Declaration.

Endnote

1   Then known as Australians for an Ecologically Sustainable Population (AESP).

References

Australian Bureau of Statistics (2013) *Population Projections, Australia, 2012 (Base) to 2101, Catalogue no. 3222.0*. ABS, Canberra.

Bureau of Meteorology (2014) *Annual Climate Statement 2013*. Australian Government, Canberra, 3 January 2014. <http://www.bom.gov.au/climate/current/annual/aus/>

Campbell CJ, Laherrère J (1998) The end of cheap oil. *Scientific American* (March), 78–83. doi:10.1038/scientificamerican0398-78

Deffeyes K (2005) *Beyond Oil – The View from Hubbert's Peak*. Hill and Wang, New York.

FAO (2011) The State of the World's Land and Water Resources for Food and Agriculture (SOLAW) – Managing Systems at Risk. Food and Agriculture Organization of the United Nations, Rome and Earthscan, London.

Grassini P, Eskridge KM, Cassman KG (2013) Distinguishing between yield advances and yield plateaus in historical crop production trends. *Nature Communications* **4**, 2918.

Heinberg R (2003) *The Party's Over: Oil, War and the Fate of Industrial Society*. New Society Publishers, Gabriola Island, BC, Canada.

International Energy Agency (2010) *World Energy Outlook 2010*. IEA, Paris.

Kunstler HJ (2005) *The Long Emergency – Surviving the Converging Catastrophes of the Twenty-First Century.* Atlantic Monthly Press, New York.

Sherwood S, Bony S, Dufresne JL (2014) Spread in model climate sensitivity traced to atmospheric convective mixing. *Nature* **505**, 37–42.

United Nations, Department of Economic and Social Affairs, Population Division (2013) *World Population Prospects: The 2012 Revision, Key Findings and Advance Tables.* Working Paper No. ESA/P/WP.227. UNDP, New York.

1

It's the numbers, stupid![1]

Paul R. Ehrlich and Anne H. Ehrlich

Humanity is faced with a daunting 'perfect storm' of environmental problems, broadly defined. The one that gets the most attention is climate disruption, which threatens the very existence of civilisation, but others may be equal or even bigger threats over the same or a longer time span. These include the loss of biodiversity and the vital ecosystem services provided to society, the depletion and destruction of rich agricultural soils, the pollution and overexploitation of surface and underground water sources, the spread of toxic synthetic chemicals from pole to pole, the deterioration of the epidemiological environment, the decline in quality and accessibility of essential mineral resources, and the prospects of even more resource wars, potentially nuclear ones.

This array is not just a list of problems, but a single interconnected complex of dilemmas, replete with ethical issues, that cannot be solved one aspect at a time. These dilemmas are all driven by a handful of factors: overpopulation worsened by continued population growth; overconsumption and consumption growth by the already rich; and the use of environmentally malign technologies, all of which are exacerbated by sociopolitical and economic inequity. These underlying drivers, of course, are not independent but strongly interrelated.

It is critical that the world's decision-makers understand this. Especially, they should not be fooled by the 'Fred Pearce Fallacy' that the critical growth issue is that of per capita consumption, not of human numbers, that the population bomb has been 'defused'. The contributions of these two factors to the human predicament can no more be separated than can the contributions of length and width to the area of a rectangle. At least several billion more people will be added before growth stops – if a crash can be avoided. But even if population growth were halted immediately, per capita consumption would still be multiplied by the gigantic number of people in the population today. That's why population *shrinkage*, now imminent or underway in some of the richest nations, is such an encouraging sign.

As the knowledgeable scientific community has repeatedly explained, the only safe course for humanity is to humanely end population growth as soon as possible and start a slow decline toward a sustainable number, reduce wasteful consumption, and focus on making key technologies less environmentally damaging. In particular, rapidly replacing fossil fuels as humanity's main energy source is essential. What is increasingly clear is that small steps and incremental change will not avert a collapse of our global civilisation. Only dramatic changes, on the scale of World War II mobilisations, hold out that hope.

The relationship of human numbers to food supply can serve as an exemplar of the complexity of these relationships. Agriculture is humanity's most important activity, and yet few citizens in heavily urbanised rich nations understand the system that nourishes them. In their view, food comes from the supermarket. They don't know, for instance, that transport is as important to that system as food growing and harvesting; that farming, fishing and processing are deeply dependent on fossil fuels; and that the food system generates roughly a third of human-derived greenhouse gas emissions. Climate disruption is an enormous threat to agricultural production even as the food system is a major cause of that disruption. Similarly, most people don't recognise that agriculture is the biggest enemy of biodiversity, although biodiversity is essential for maintaining food production through pollination, pest control and the maintenance of soil fertility. And the consequences of long-term trends of soil deterioration and groundwater overexploitation are potentially horrendous.

When we wrote *The Population Bomb* in 1968 (Ehrlich 1968), the global population was 3.5 billion. We were heavily criticised for taking a pessimistic view of world food prospects and for claiming that 'the battle to feed all of humanity has been lost'. Various critics explained that wonderful technological advances would allow humanity to feed and otherwise care for five, 10, or even 16 billion people. Of course, that hasn't happened. At the moment nearly a billion human beings are seriously underfed, with perhaps five million dying annually from starvation and hunger-related diseases. Our view today remains the same as in 1968: Let's show we can develop a production and distribution system that successfully feeds today's population; until then it is highly unethical to babble about how it will be possible in the future to feed a much larger population.

If food were in some sense 'fairly' distributed now, everyone theoretically could be adequately fed. But economic inequity rears its ugly head here; not only have human beings never distributed food fairly, there is no sign they are about to do it. Furthermore, demographers estimate that 2.5 billion mouths (many more people than were alive when we were born) are likely to be added to the population by 2050, even though there are already signs of climatic effects on agriculture that might so reduce food production that everyone could not be fed even if food were distributed fairly.

The challenge to agriculture alone is monumental. The nonlinearities in the food system are almost never discussed. Human beings are smart, and our ancestors naturally picked the low-hanging fruit first. They didn't start farming on marginal lands first and then gradually move to the rich soils of river bottoms. Even starting with the richest areas, in many cases, the nutrients in those soils have become depleted, and many productive areas have been paved over for urban development. Moreover, pioneering agriculturalists started farming where rainfall was sufficient to support crop growth, and early on began to divert water from river systems to irrigate land where rainfall was inadequate. When farmers began using metal instruments in farming, they had access to high-grade ores which now have been seriously depleted. So today, on average, each new person added to the population must be fed from poorer soils, supplied with water pumped from deeper wells, transported further, or purified more – all involving increased energy use. Whereas early farmers in some places had nuggets of nearly pure copper available on or just under Earth's surface; today in some places copper is being mined at depths of over a mile and from ores with copper concentrations of only 0.5 per cent.

Scientific assessments of how to support an expected explosion of human numbers from 7.1 to 9.6 billion by 2050, to say nothing of 10–11 billion in 2100, all too often complacently accept that much population growth as a given. The notion that a dramatic campaign to increase women's rights and opportunities and supply modern contraception and safe backup abortion to all sexually active human beings might instead lead to a slowly shrinking population of 8.5 billion in 2050 is rarely considered. Having a billion fewer people to feed and supply with goods a few decades from now could greatly reduce the odds of vast famines.

The role of human numbers in disrupting the climate and increasing the chances of famine is obvious. Other parts of the perfect storm have large population components as well, though sadly little recognised. Even a quarter century ago, when the global population was only about five billion, *Homo sapiens* was already using, co-opting or destroying some 40 per cent of all plant growth on land – that is, the basic food supply of all animals. That is a measure of the impact of the human enterprise on Earth's precious biodiversity.

The more people there are, *ceteris paribus*, the more toxic substances are introduced into the environment. Especially serious are hormone-disrupting compounds that can have serious effects on animals' and human development at extremely low concentrations. The use of many toxic materials in agriculture can contribute to biodiversity losses (pollinators, pest-controllers, soil microbes), which in turn have negative impacts on crop yields.

It is also well known that the size, density and mobility of the human population are crucial factors in the odds of vast epidemics. Our infectious diseases all originated in other animals. The more people there are, and the more that are malnourished, the greater are the chances of a nasty pathogen transferring into a human population, and the greater the chances that it will persist and spread.

Finally, it is increasingly obvious that growing human numbers are generally forcing humanity to exploit poorer and less accessible resources under more and more dangerous circumstances. This can clearly be seen in the history of oil exploration. The first oil well in the United States, in Pennsylvania (1859), drilled down 21.2 m below the surface to strike oil. The Deep Water Horizon's well, which caused the great Gulf of Mexico oil spill in 2005, descended through 1.6 km of water and went down 3962.4 m further. That's an example of the nonlinearities in the resource system, dramatically showing diminishing returns to complexity, which is considered one important element in the collapse of complex societies. But the Deep Water Horizon well is only an extreme case; such constraints increasingly hinder the energy mobilisation industry everywhere, causing a general decline in the energy return on energy invested (EROEI).

Similar challenges are confronting the mining industry as well, as declining resource quality demands more energy (and often more water) to extract minerals. Few people recognise the gigantic scale of mining operations today. For example, Australia exports enough coal annually to form a pile 1 m high and 20 m wide circling the entire Earth. And the country 'plans' to increase that several-fold, with operations that will inevitably attempt to move disproportionately more rock (with disproportionate environmental damage) in order to achieve economies of scale. As Simon Michaux has pointed out in detail, mineral extraction at current rates is not sustainable – a situation with profound consequences for standard ideas of sustainability (Michaux 2011).

The biophysical problems presented to humanity because of its overpopulation have repeatedly been pointed out unambiguously by the scientific community. Yet the sociopolitical consequences of having too many people, and the profound ethical issues of not caring for the situation of our descendants, are less often discussed. Crowding and congestion are easily recognised as related to population numbers (how could they not be?), but amazingly the impact of numbers on political systems is rarely mentioned, even though it concerned James Madison more than two centuries ago. In the Federalist-anti-Federalist debates, Madison suggested that one congressman (all were male then) might reasonably represent 30 000 people, and wondered how that would be modified as the population of the United States expanded. Today each congressperson represents on average more than 900 000 people. More people clearly mean less democracy as well as less natural capital (including waste sinks) to serve our descendants. Is it ethical, then, for anyone to have more than one or two children?

What can be done to increase the chances of avoiding a catastrophic collapse of civilisation? First, of course, it is necessary to initiate an effective program to humanely stop population growth and begin a slow decline. The formula for doing this is simple. First, educate girls and give women everywhere the rights, opportunities and financial rewards now available to men. Second, provide every sexually active human being with access to modern contraception and backup abortion. Given adequate support, this plan might end growth in a few decades and lead to a gradual shrinkage of the population. Of course, taking those steps would require monumental social change in the societies where large families still prevail. Few believe success can be achieved in time against the opposition of traditional norms and religious and business interests with strong pro-natalist incentives, financial stakes and no concern for future generations.

There may be more hope for success with the other major driving force causing lethal environmental deterioration: overconsumption by the rich. Here we can take heart from the record of mobilisations for World War II. In 1942 the entire United States economy was transformed, rationing of many commodities was instituted, millions of men were drafted into the military and millions of women and African-Americans joined the industrial labour force. In 1946 the entire process was reversed (except for the equity advances, which were not entirely abandoned). Thus, with appropriate incentives, huge changes in consumption patterns could be generated.

But what incentives are there to push humanity into taking the giant collective steps required to save civilisation? Most people would doubtless be willing to take dramatic action if they understood the situation and what was needed. But most people are utterly clueless, thanks to a general failure of education systems and the media to inform the public appropriately. The failure is not accidental but a function of the short-term interests of those who own the media and own the politicians. The political dilemma is doubtless more critical; indeed we believe that getting the money out of politics is the greatest ethical challenge we face.

We think the best hope for starting the necessary cultural revolution is to enhance and coordinate the activities of the multitude of NGOs in civil society – and that is a major goal of the *Millennium Alliance for Humanity and the Biosphere* (MAHB <mahb.stanford.edu>). Joining costs nothing and takes only a few minutes. Get involved and you may be able to contribute to making civilisation sustainable for at least a while.

Endnote

1   More detail and references on the technical issues can be found in Ehrlich and Ehrlich (2013).

References

Ehrlich PR (1968) *The Population Bomb*. Ballantine, New York.

Ehrlich PR, Ehrlich AH (2013) Can a collapse of global civilization be avoided? *Proceedings of the Royal Society B* **280**, 20122845.

Michaux SP (2011) Three caveats that will change our design culture. *AusIMM Bulletin* **6**, 66–71.

2

The environmental implications of population growth

David Lindenmayer

One of Paul Ehrlich's early studies relates environmental impact to population size and per capita consumption (that is, Impact = Population × Affluence/consumption × Technology, or I = PAT). A very important aspect of this is T, or technology, which corresponds to the environmental damage that occurs as a result of supplying each unit of consumption. A classic recent paper by Davidson and Andrews (2013) examined issues beyond population size and the level of consumption and explored impacts of per unit consumption. They highlighted issues associated with ecosystem condition resulting from resource use. For example, biodiversity loss can be an outcome of the impacts of resource extraction and consumption and it needs to be captured in amended versions of the IPAT equation outlined above. Indeed, I argue that the status of biodiversity is a crude, but nevertheless useful, indicator of the ecological sustainability (or otherwise) of resource extraction and consumption, and hence the ecological sustainability of human populations.

A key consideration in the ecological sustainability of the human population is that most kinds of natural resources are now significantly harder to extract than they were previously. In his chapter on mining in the book *Ten Commitments Revisited*, Mudd (2014) discusses how the concentration of ore bodies is declining

Figure 2.1: Open-cut mine. (Photo: Matthew Roberts.)

dramatically in Australia and elsewhere. Therefore, the extraction of resources now has the potential to lead to far greater environmental impacts than it did, for example, two or more decades ago. What this means is that even if population size remains constant, and levels of consumption remain unchanged, increasingly difficult resource extraction leaves a significantly greater environmental footprint with increased effects on other aspects of the environment, like biodiversity. To further illustrate this issue, Mudd (2014) estimates that, in Australia, mining is now producing billions of tonnes of waste rock and tailings annually. Open-cut mining may therefore be producing degraded areas of the environment that are larger in size than areas we are able to restore and remediate effectively, resulting in negative impacts on other aspects of the environment such as the availability of suitable habitats for native plants and animals.

In another example, there has been much recent discussion about the impacts of coal seam gas on aquifer recharge in the Murray–Darling Basin, and potentially also elsewhere in the Great Artesian Basin. Yet the extent of clearing and fragmentation of terrestrial vegetation, the increasing populations of weeds and feral animals and other impacts of coal-seam gas exploration and extraction have rarely been discussed.

Population growth in Australia

The rate of population growth in Australia is higher than many other countries in the OECD – as are levels of resource use and per capita energy consumption. When some politicians argue for a substantially 'bigger Australia', such as suggesting the doubling of current population size by 2050, they must concurrently consider the environmental implications of such large and rapid increases in population size. For example, such increases would translate to another 65 cities the size of Canberra, or Melbourne and Sydney becoming mega-cities beyond eight to 10 million people (see Foran and Poldy 2002). A lot of that expansion would take place in threatened or endangered environments where there are already large numbers of threatened species. Take, for example, the two suburbs of Wright and Coombs that are expanding on the urban fringe of Canberra into nationally endangered temperate woodland environments.

There are additional environmental major impacts that arise from large and rapid increases in human population size. As examples: (1) Based on figures calculated in 2002, around 200 tonnes of natural resources must be moved each year to maintain each Australian at current standards of living – 2.5 times higher than for a person in the USA and five times greater than for an individual in Japan. The environmental impacts of additional resource use and movement associated with each extra person have been outlined above. (2) Every extra person in Australia is responsible for about an additional 24 tonnes of CO_2 emissions per capita, per year. That's twice the OECD average and four times the world average (Hughes 2014). The environmental impacts of additional greenhouse gas emissions are well documented, including on Australian biodiversity (Steffen *et al.* 2009). (3) Every additional Australian resident costs another \$340 000 extra support in terms of built assets. Thus, if 600 000 people are added to the population of Western Sydney in the coming two decades, this will translate to an extra \$204 billion of additional infrastructure costs required for those people – on top of provision for existing residents in the area (Sobels and Foran 2014). Extra funding for built infrastructure could prevent us dealing with other things in the environment, including many environmental problems that we certainly have not tackled particularly well to date, such as biodiversity conservation.

Population size, agricultural intensification and environmental impacts

Yet other environmental issues arise from continued human population growth in Australia. One of these is indirect and it entails poorly conceived notions that Australia will become 'a food bowl for Asia'. Many investigations over many years, including extensive reports by the former CSIRO Division of Land Use Research, have indicated that the potential to markedly increase food production in northern

Australia is very limited at best (see also Lindenmayer *et al.* 2014). Similarly, there is likely to be only limited potential to increase food production in southern Australia, as many ecosystems are unlikely to sustain agricultural intensification without further soil erosion and secondary salinity, clearing of marginal land and losses of biodiversity (Cunningham *et al.* 2013). Therefore, Australian responses to increasing global human populations by significantly increasing export food production could not only result in the depletion of the nation's environment, but also eventually be counterproductive through degrading important food and fibre producing ecosystems (Lindenmayer *et al.* 2012).

The education–environment paradox

There is an extraordinary paradox in Australia between our perception of the human population, levels of education and environmental issues. Ours is one of the richest nations, biologically, in the world. It is also one of the most environmentally degraded nations in the world. Australians are relatively well educated, but we are not well educated in issues associated with human population size and long-term ecological sustainability. I emphasise the word 'ecological' here, as a part of true sustainability. This is because, to be meaningful, it must include, for example, the status of biodiversity conservation as an indicator of long-term ecological sustainability. Often the concept of sustainability is abused because its environmental context is ignored. Indeed, without such a context, sustainability becomes what is termed a *panchreston* (Lindenmayer and Fischer 2007). That is, a catch-all term that means all things to all people but lacks substance or specific meaning. This leads to its misuse – such as 'sustainable' coal industries, and 'sustainable' other kinds of fossil fuel industries, but which are demonstrably unsustainable in an ecological sense.

A more robust debate about population size and ecological sustainability is well overdue in Australia. These debates also must canvass issues about the relationships between economic growth, population size, levels of resource consumption, the environmental impacts of (increasingly difficult) resource extraction, environmental degradation and the loss of biodiversity as an indicator of that degradation. Population debates need to be separated from other issues like asylum seeking, and racism and religious preferences, as these latter issues often derail attempts at rational discussion and evidence-based policy. Politicians and policy-makers also need to be far better engaged with population issues – the former Federal Department of Sustainability, Environment, Water, Population and Communities contributed little to discussions about the environmental impacts associated with different scenarios for the Australian population. These debates (and the hopefully well-informed policies that derive from them) are crucial as they will shape what this country will look like in the coming decades and determine whether our nation is on an ecologically sustainable path or something environmentally disastrous.

References

Cunningham SA, Attwood SJ, Bawa KS, Benton TG, Broadhurst LM, Didham RK, McIntyre S, Perfecto I, Samways MJ, Tscharntke T, Vandermeer J, Villard MA, Young AG, Lindenmayer DB (2013) To close the yield-gap while saving biodiversity will require multiple locally relevant strategies. *Agriculture, Ecosystems & Environment* **173**, 20–27.

Davidson DJ, Andrews J (2013) Not all about consumption. *Science* **339**, 1286–1287.

Foran B, Poldy F (2002) *Future Dilemmas: Options to 2050 for Australia's Population, Technology, Resources and Environment: Report to the Department of Immigration, Multicultural and Indigenous Affairs.* CSIRO Sustainable Ecosystems, Canberra.

Hughes L (2014) Climate change. In *Ten Commitments Revisited: Securing Australia's Future Environment*. (Eds D Lindenmayer, S Dovers and S Morton) pp. 217–225. CSIRO Publishing, Melbourne.

Lindenmayer D, Fischer J (2007) Tackling the habitat fragmentation panchreston. *Trends in Ecology & Evolution* **22**, 127–132.

Lindenmayer D, Cunningham S, Young A (2012). *Land Use Intensification – Effects on Agriculture, Biodiversity and Ecological Processes*. CSIRO Publishing, Australia.

Lindenmayer D, Dovers S, Morton S (Eds) (2014) *Ten Commitments Revisited: Securing Australia's Future Environment*. CSIRO Publishing, Melbourne.

Mudd GM (2014) Mining. In *Ten Commitments Revisited: Securing Australia's Future Environment*. (Eds D Lindenmayer, S Dovers and S Morton) pp. 157–164. CSIRO Publishing, Melbourne.

Sobels J, Foran B (2014) Population. In *Ten Commitments Revisited: Securing Australia's Future Environment*. (Eds D Lindenmayer, S Dovers and S Morton) pp. 263–270. CSIRO Publishing, Melbourne.

Steffen W, Burbidge A, Hughes L, Kitching R, Lindenmayer DB, Musgrave W, Stafford-Smith M, Werner P (2009) *Australia's Biodiversity and Climate Change*. CSIRO Publishing, Melbourne.

3

Whither wildlife in an overpopulated world?

Chris R. Dickman

Using existing crop plants and methods it may not be practicable to produce adequate food for more than four doublings of the world population, though the complete elimination of all land wildlife ... might allow two or three further doublings. (J.H. Fremlin, 1964)

Introduction

In the 50 years since Fremlin explored technical limits to the number of people that the world can support, the human population has burgeoned from three billion to more than seven billion people. The time taken for the population to double has increased slightly from the 37 years that Fremlin used as the basis for his calculations but, currently doubling every 50 years, the world population is clearly still on a fast upward trajectory. Rates of growth are generally higher in developing compared with developed parts of the world and are expected to contribute most to the increase in population size in future. Using different assumptions about human fertility and life expectancy, the Population Division of the United Nations predicts that the world population at the end of the 21st century could be anywhere from 6.8 billion to 28.6 billion people.[1] Notably, the Population Division of the United Nations does not comment on whether the world might have sufficient resources to continue growing, nor does it have anything to say about the impacts of continued human population expansion on other forms of life.

In this chapter, I discuss the general relationship between human population growth and wildlife in Australia. On the one hand, human population density on the island continent is relatively low – just three people/km^2 compared with the world average of 53 people/km^2 – suggesting perhaps that the impacts of humans on native Australian wildlife should be correspondingly small. On the other hand, Australia's wildlife is mostly endemic and, having evolved for 45 million years in isolation from the rest of the world, is sometimes seen as being more 'fragile' and less resistant to the effects of new species than wildlife elsewhere (Dickman 2007). I will show that native Australian wildlife has indeed fared poorly in the past in the presence of expanding human populations, and that many of the continent's most iconic and charismatic species will be at risk if current trends continue. I will conclude by arguing that unless strenuous efforts are made now to stabilise human numbers and impacts, a dystopian future will be hard to avoid.

Australia's human population

The first Australians arrived at least 50 000 years ago and spread to all parts of the continent. The size of the Indigenous population undoubtedly varied over this period, increasing initially as small numbers of colonists became established and then fluctuating in correspondence with shifts in the climate and the prevailing resource base. Although debated, the Indigenous population probably did not exceed one million at any time. In 1788, new colonists arrived, at first as a trickle, but then in very rapidly increasing numbers. By 1900 the national population was around 3.8 million, by 1950 it was 8.3 million and just 50 years later it was 19.3 million.[2] By October 2013 the population exceeded 23 million.

Projections made by the Population Division of the United Nations indicate that Australia can expect its population to continue to increase in most scenarios, perhaps achieving 41.5 million under 'medium variant' assumptions by 2100 (Figure 3.1). Projections made by the Australian Bureau of Statistics, however, which take greater account of the nation's extraordinarily high rate of overseas net migration, suggest that the population may reach between 42.4 million and 70.1 million people in 2101.[3] Whatever the actual number, Australia looks as if it is likely to achieve at least two and perhaps three doublings in population size from the time of Fremlin's essay until the end of the 21st century. Remember that

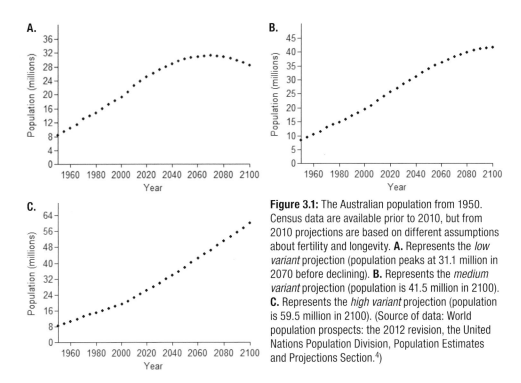

Figure 3.1: The Australian population from 1950. Census data are available prior to 2010, but from 2010 projections are based on different assumptions about fertility and longevity. **A.** Represents the *low variant* projection (population peaks at 31.1 million in 2070 before declining). **B.** Represents the *medium variant* projection (population is 41.5 million in 2100). **C.** Represents the *high variant* projection (population is 59.5 million in 2100). (Source of data: World population prospects: the 2012 revision, the United Nations Population Division, Population Estimates and Projections Section.[4])

Fremlin expected the fourth doubling to herald the beginning of the apocalypse for all land-based wildlife: how can we expect wildlife to fare before we reach such a grim scenario?

People and wildlife in Australia – pre-1788

The first people to arrive in Australia would likely have confronted a bestiary that they had not encountered before. Among the new creatures would have been giant lizards, snakes, flightless birds and marsupials. Collectively, these now-extinct behemoths have been termed *megafauna*. The demise of these large animals began at about the same time, or just after, the arrival of the first waves of human immigrants, leading many researchers to argue that people caused the extinction event. Although other explanations have been put forward to explain the loss of the megafauna, such as climate change, an increase in the frequency and intensity of fire, and disease, the hypothesis of rapid overkill by humans is compelling. This follows from the very slow rate of reproduction of large animals: even the occasional hunting of young animals could have driven population replacement rates of susceptible species into negative territory, causing them to disappear within just a few hundred years of their initial contact with humans (Johnson 2006). If this interpretation is correct, the arrival of the first people was a harbinger of other, more deadly, events to come. The first of these events was the introduction of the dingo (*Canis dingo*); the second was the hugely disruptive arrival of white settlers in 1788.

The ancestors of the modern dingo were introduced at least 4000–4500 years ago, almost certainly by Asian seafarers. Their subsequent spread across the continental mainland has been linked to the losses of a flightless native hen (*Gallinula mortierii*), the Tasmanian devil (*Sarcophilus harrisii*) and thylacine (*Thylacinus cynocephalus*), all of which persisted on the dingo-free island of Tasmania. However, predation by – and perhaps competition from – the dingo does not fully account for the demise of these three species on the continental mainland. Around the time of the introduction of the dingo, the mainland human population was increasing and people's hunting efficiency was improving due to the adoption of new weaponry such as edge-ground and hafted stone tools. These changes affected Tasmania to a much lesser degree. As a result, it seems most likely that the demise of the hen, the devil and the thylacine on mainland Australia resulted from a combination of direct human hunting pressure as well as impact from a novel canid predator that people had introduced.

People and wildlife in Australia – 1788 to the present

Waves of immigration by Britons, continental Europeans and then other nationalities from the late 18th century set in train suites of processes that have had by far the most profound and dramatic effects on Australia's native wildlife. Most species have fared poorly following the arrival of the new Australians, but a few have benefited from access to new habitats and resources that have been created. Considering just terrestrial vertebrates, 54 species and subspecies are listed as extinct on the schedules of the *Environment Protection and Biodiversity Conservation Act 1999*, and a further 290 species are considered to be at some risk of extinction in future.[5] All the listed species have declined or faded from existence since 1788. In addition to the vertebrates, the habitats that support many species are at risk, with a recent analysis listing some 3000 'ecosystem types' that are in this category.

Before interpreting these statistics, we need to address two questions. First, is the documented extinction rate higher than the natural, or background, extinction rate? Second, if so, to what extent have humans contributed? In a review of paleontological and molecular studies, Pimm *et al.* (1995) calculated that species-longevity is in the order of a million years, sometimes a little more. It follows from this that in any given year we might expect one species in a million to expire from natural causes. Taking Australian terrestrial mammals and a baseline date of 1788, the loss of 20 species from an initial total of 312 represents a loss rate of ~285 extinctions per million species per year. Similar extrapolations for native frogs and birds yield loss rates of

~82 and ~57 extinctions per million species per year, respectively; no reptiles are known to have become extinct since 1788. These simple estimates confirm that the actual rates of recent extinction of mammals, birds and frogs are many times greater than background rates. In the section below, I consider the evidence that humans have directly or indirectly caused the post-1788 extinctions.

Direct effects of people on wildlife

At different times, most jurisdictions in Australia have declared war on wildlife or on particular target species. In Queensland, for example, the parliament passed *An Act to Facilitate and Encourage the Destruction of Marsupial Animals* in 1877. When this Act and its successors were replaced in 1930, over 27 million mammals – mostly marsupials – were documented to have been killed, although the true number was almost certainly much greater (Hrdina 1997). Many millions of native mammals, birds and reptiles were killed over the same period in New South Wales under similar legislation, with very large numbers poisoned by the widespread and indiscriminate laying of strychnine baits (Dickman 2007; G. Bernardi personal communication). Rats were the subject of a concerted culling campaign in Sydney following an outbreak of bubonic plague in 1900; bettongs, bandicoots, quolls and other taxa were targeted for local extermination in campaigns elsewhere, with at least some of these campaigns achieving success (Dickman 2007). The most high-profile victim of these early pogroms was undoubtedly the thylacine, which succumbed to hunting pressure and perhaps an epidemic disease early in the 20th century. Culling of threatened species still occurs (e.g. the grey-headed flying-fox, *Pteropus poliocephalus*), ostensibly only under licence.

Indirect effects of people on wildlife

The arrival of new settlers from 1788 resulted in many changes to the Australian environment, and these have had indirect but pervasive effects on wildlife. It is difficult now to discern the relative importance of these changes, and it is likely that they have affected different components of the vertebrate fauna in different places at different times. Nonetheless, there is general agreement that habitat change and the introduction of invasive herbivores and predators have had particularly damaging effects, and these are considered briefly below. This topic has been much-discussed, and detailed reviews – among many that have been written – can be found in Rolls (1969), Johnson (2006) and Dickman (2007).

In the first instance, the fragmentation and loss of habitat reduce food and shelter resources, resulting in the decline or disappearance of species that require them. For example, the conversion of almost all native grassland for agriculture and urban infrastructure in many regions has reduced the numbers of grassland-dependent species and has led to local or regional extinctions of several (e.g. the eastern barred bandicoot, *Perameles gunnii*, in Victoria; striped legless lizard, *Delma impar*, in much of south-eastern Australia). The grazing of livestock in coastal, sub-coastal and inland areas has depleted or destroyed vast areas of shrubland and woodland and contributes to associated problems such as soil erosion, salination and the loss of natural springs and wetlands. The livestock industry also drove the creation of the notorious Marsupial Destruction Acts that were noted above. In reviewing the factors that led to the collapse of ground-dwelling native mammals in western New South Wales, Lunney (2001) laid the blame squarely with the excesses of the sheep industry around the turn of the 20th century. Other production industries such as silviculture, woodchipping and sawlog harvesting affect many arboreal species such as forest birds, bats and marsupials.

Broadscale production industries have obvious potential to affect wildlife habitat, but more locally focused activities such as road-building, mining, coal seam and shale gas extraction (fracking) can also affect wildlife populations. Taking reptiles as an example, Wotherspoon and Burgin (2011) estimated that between 4576 and 6838 individuals were killed annually on minor roads within a handful of towns and villages in the Blue Mountains in New South Wales, and labelled this loss the 'tip of the iceberg … an ecological disaster'. Near Coober Pedy, in South Australia, Pedler (2010) estimated that between 10 million and 28 million reptiles are killed annually by falling into shafts that have been left uncapped by opal prospectors. All urban, production and industrial enterprises are accompanied by further changes that affect wildlife. These include contamination of the air, water and soil by organic and inorganic materials, and subtle but pervasive pollution effects of increased noise and light.

The second major category of human-caused indirect effect on wildlife is the introduction of non-native species that then impact upon their native counterparts. In Australia, such species include the European rabbit (*Oryctolagus cuniculus*), red fox (*Vulpes vulpes*), house cat (*Felis catus*), black rat (*Rattus rattus*), pig (*Sus scrofa*), goat (*Capra hircus*) and six species of deer; Olsen (1998) listed a further 30 species of terrestrial vertebrates that could be considered pests, as well as 13 species of fish. The impacts of these invasive species are classically considered to be wrought via competition for food or shelter resources or, in the case of the carnivorous fox and cat, via predation. However, cats host parasites such as *Toxoplasma gondii* and *Spirometra erinacei* that can detrimentally affect native fauna via disease, while the cane toad (*Rhinella marina*) is notorious for producing toxins that rapidly poison native predators that attempt to eat them.

Invasive vertebrates have achieved most prominence for their impacts on native wildlife, but other species imported by humans can also cause damage. For example, the yellow crazy ant (*Anoplolepis gracilipes*) was first reported on Christmas Island in 1934 and appears to be driving a widespread ecological collapse of the island's ecosystems. Most endemic vertebrates are declining, with the Christmas Island pipistrelle (*Pipistrellus murrayi*) and shrew (*Crocidura trichura*) the most recent species to succumb (Eldridge *et al.* 2014). The crazy ant has been implicated in the extinctions of both species. Environmental weeds that were introduced originally to improve pastures, stabilise soil or to augment the horticultural industry, or species that hitched rides on boats or aeroplanes, also cause problems for wildlife in many areas by replacing natural food or shelter resources.

A further indirect effect of human activity that can be expected to affect wildlife is climate change. Bioclimatic modelling suggests that areas that currently provide many species with a suitable climate will shift in decades to come, so that persistence may be assured only if these taxa are able to track and keep within their preferred climatic 'envelope'. This will be difficult for species that now occur in high altitude or high latitude regions as the cool conditions that now prevail will disappear in future. Migratory movements will also be difficult or impossible for sedentary species or for those whose habitats have been fragmented and thus have no corridors along which to move. More subtle effects of climate change can be expected too. For example, many species will need to access sheltered refuge sites to withstand the intense wildfires and more-frequent droughts and heatwaves that are expected in the southerly regions of Australia. Crowther *et al.* (2014) have pointed out recently that even widely distributed species such as the koala (*Phascolarctos cinereus*) will be at grave risk if mature shade trees are not available to afford protection.

The factors that influence wildlife may act alone, but in most situations are much more likely to act together in additive or interactive ways to bring about their effects. Thus, the introduction and establishment of the European rabbit probably elevated populations of the red fox and feral cat, in turn allowing these predators to exert stronger downward pressure on populations of susceptible ground-active birds and marsupials. The removal of shrubs and trees to improve pastures for domestic grazing animals would have reduced the shelter that was available to native species, exacerbating the hunting opportunities for the novel predators still further. The introduction of new fire-dependent savannah grasses to northern Australia allows hotter and more widespread fires to burn, again simplifying the habitat and providing more open hunting areas for feral cats. Examples of such synergisms are legion, and the effects they have on many components of native wildlife can be catastrophic.

We can conclude that, overall, people have had dramatically negative direct, indirect and synergistic effects on native Australian wildlife. A few species have done well, including those that are tolerant of disturbance and the presence of people (e.g. some corvids, noisy miners *Manorina melanocephala*, rainbow lorikeets *Trichoglossus haematodus*), or those that can exploit new habitats and food sources such as improved pastures (e.g. some of the large kangaroos), but most species have fared poorly. What can we expect if Australia's human population continues to increase?

Back to the future

Making predictions about the future is fraught with difficulty and, of course, can be judged only with the benefit of hindsight. Even Nostradamus got things wrong! With this in mind, in this section I will use records on the timing of species' extinctions that go back to 1950 (the recording of extinction events was generally poor before this) to make some extrapolations of what might occur in the future. By plotting the date that a species was last recorded against the human population that prevailed at that time, we can then look for a relationship between the two variables. The resulting plot (Figure 3.2) is quite revealing.

It shows, most obviously, a positive relationship between the number of people in Australia and the cumulative number of extinctions of terrestrial vertebrates. This is inevitable: extinctions have occurred and are irreversible, and the human population has increased monotonically over the last several decades. However, the tightness of the relationship is surprising, with 96 per cent of the variation in the cumulative number of extinctions being explained by human population size. The regression equation indicates that almost one (0.95) vertebrate species has become extinct with every additional million people since 1950.

If we assume that this relationship will hold in future, we can use the regression equation to extrapolate how many more species may disappear. Thus, using the Australian Bureau of Statistics projections noted above, by 2101, 21 more native vertebrates could be extinct if the human population reaches 42.4 million, and a further 26 species (i.e. 47 new extinctions) if the population reaches 70.1 million. If these extinctions were to occur, with at least 56 species having disappeared already since 1788, Australia's tally of extinct vertebrates by 2101 would be 77–103 species. At most risk would be charismatic species such as upland frogs, many woodland birds and reptiles, island endemics, and large and iconic species such as the koala, some quolls and remaining species of bandicoots and rat-kangaroos. Such losses would dramatically worsen what is already the world's highest rate of mammalian extinction and elevate losses of other vertebrates to distressingly high levels too.

Such extrapolations are, of course, nothing more than formal projections of current trends, and there is no guarantee that the threats unleashed by humans on wildlife in the past will be the same as those to come. For example, feral cats and red foxes will arguably have less dramatic effects in future because they have already driven the most susceptible species to extinction; climate change, conversely, will probably have increasingly devastating effects.

Despite such uncertainties, there is no doubt that more people mean more impact, more species driven to extinction, and more species sliding towards the abyss. Although I have focused on the more obvious threats so far, there is a further insidious threat that should be noted. This is the problem of cultural memory loss. As we become increasingly urbanised and more of us arrive in the

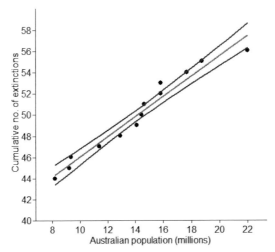

Figure 3.2: The relationship between human population size in Australia and the cumulative number of extinctions of terrestrial vertebrate species. Between 1950, when the Australian population was 8.3 million, and 2009 when the population had reached 22 million, four species of native frogs and nine species of mammals had become extinct. Extinction was taken as the date of the last record of these species from information provided in the schedules of the *Environment Protection and Biodiversity Conservation Act (EPBC) 1999*.[6] Two further species, not listed as extinct in the EPBC Act, were included: the Christmas Island shrew (*Crocidura trichura*), last seen in 1985, and the Christmas Island pipistrelle (*Pipistrellus murrayi*), last seen in 2009 (Eldridge *et al.* 2014). Before 1950 there were 43 species of terrestrial vertebrates that became extinct at various times. The graph shows the 13 species that disappeared between 1950 and 2009 (dots), the linear regression (straight line) and 95% confidence intervals (curved lines). The equation for the regression is $y = 36.5 + 0.95x$, $R^2 = 0.96$. For every million people added to the Australian population since 1950, almost one (0.95) species has become extinct.

country with no Australian sense of place, our connection to the continent's natural environment collectively diminishes. We then become less concerned when the environment and its natural inhabitants are degraded or destroyed; why should we worry when there has been little or no appreciation of the continent's natural riches in the first place? Three of the most environmentally hostile state governments that have been seen in recent decades have just been elected in eastern Australia. Do the 'open for (extractive, exploitative) business' models of these governments reflect the collective sentiments of people who are increasingly disengaged from the natural environment? If this interpretation is correct, we are setting up a positive feedback loop that will allow accelerating loss of wildlife and other species while we look the other way.

Conclusions

The inexorable increase in human population size in recent decades in Australia and elsewhere has been accompanied by dramatic reductions in the population sizes of many non-human species and by the extinction of many others. Humans can be implicated directly and indirectly in the cascade of extinctions so far, and ever-increasing numbers of people will inevitably extirpate more species, the habitats they use and the ecological processes and services they provide. Some losses initially may be mourned, but most will pass unnoticed and almost all seem destined to fade from our collective memory. Can we avoid Fremlin's wildlife apocalypse and a dystopia Down Under? Perhaps, but strenuous efforts must be made now to stabilise human numbers and impacts. This in turn will require debate at the national and international levels, and more forward thinking and moral courage than we have collectively shown so far.

Acknowledgements

I thank Jenny Goldie and Hugh Tyndale-Biscoe for the opportunity to take part in the Fenner Conference, many participants at the meeting who sharpened my appreciation of the problems engendered by the problem of human overpopulation, and Carol McKechnie and Alice Dickman for their support and critical reading of the manuscript.

Further reading

Baird V (2010) Too many people? *New Internationalist* **429**, 4–20.
O'Connor M, Lines WJ (2010) *Overloading Australia: how Governments and Media Dither and Deny on Population*, 4th edition. Envirobook, Sydney.

Endnotes

1   See: <http://esa.un.org/unpd/wpp/unpp/panel_population.htm>, accessed 1 October 2013.
2   See: <http://www.abs.gov.au/AUSSTAYS/abs@.nsf/DetailsPage/3105.0.65.0012008?OpenDocument>, accessed 1 October 2013.
3   See: <http://www.abs.gov.au/ausstats/abs@nsf/Lookup/3222.0main+features52012%20(base)%20to%202101>, accessed 2 October 2013.
4   Available at: <http://esa.un.org/wpp/unpp/panel_population.htm>, accessed 1 October 2013.
5   See: <http://www.environment.gov.au/cgi-bib/sprat/public/publicthreatebedlist.pl>, accessed 1 October 2013.
6   See: <http://www.environment.gov.au/cgi-bin/sprat/publicthreatenedlist.pl>, accessed 2 October 2013.

References

Crowther MS, Lunney D, Lemon J, Stalenberg E, Wheeler R, Madani G, Ross KA, Ellis M (2014) Climate-mediated habitat selection in an arboreal folivore. *Ecography* **37**, 336–346.

Dickman CR (2007) *A Fragile Balance: The Extraordinary Story of Australian Marsupials*. Mallon Publishing, Melbourne.

Eldridge MDB, Meek PD, Johnson RN (2014) Taxonomic uncertainty and the loss of biodiversity on Christmas Island, Indian Ocean. *Conservation Biology* **28**, 572–579.

Fremlin JH (1964) How many people can the world support? *New Scientist* **24**, 285–287.

Hrdina FC (1997) Marsupial destruction in Queensland 1877–1930. *Australian Zoologist* **30**, 272–286.

Johnson C (2006) *Australia's Mammal Extinctions: a 50 000 Year History*. Cambridge University Press, Cambridge.

Lunney D (2001) Causes of the extinction of native mammals of the Western Division of New South Wales: an ecological interpretation of the nineteenth century historical record. *The Rangeland Journal* **23**, 44–70.

Olsen P (1998) *Australia's Pest Animals: New Solutions to Old Problems*. Bureau of Resource Sciences, Canberra, and Kangaroo Press, Sydney.

Pedler RD (2010) The impacts of abandoned mine shafts: fauna entrapment in opal prospecting shafts at Coober Pedy, South Australia. *Ecological Management & Restoration* **11**, 36–42.

Pimm SL, Russell GJ, Gittleman JL, Brooks TM (1995) The future of biodiversity. *Science* **269**, 347–350.

Rolls EC (1969) *They All Ran Wild: the Story of Pests on the Land in Australia*. Angus & Robertson, Sydney.

Wotherspoon D, Burgin S (2011) The impact on native herpetofauna due to traffic collision at the interface between a suburban area and the Greater Blue Mountains World Heritage Area: an ecological disaster? *Australian Zoologist* **35**, 1040–1046.

4

The outlook for population growth in Australia

Bob Birrell

The purpose of this chapter is to provide a realistic analysis of the population outlook for Australia. The core concern is to identify the key political and economic factors which are currently taking Australia down a high-growth path, despite the concerns of those worried about the environmental and quality of life consequences of proceeding down this pathway. Of course, things could change should these concerns be mobilised in the political arena. At present, however, the analysis suggests the prospects for this happening in the near future are not bright.

Table 4.1 provides the most recent estimates of Australia's resident population growth and the role of natural increase and net overseas migration (NOM) in fuelling this growth. In 2012, Australia's population grew by 1.8 per cent. This is higher than in any other developed country. It is evident that NOM is the main driver. This will remain the case because the scale of natural increase (births minus deaths) and the rate of growth attributable to natural increase will slowly fall should Australia's fertility rate remain a little below replacement level.

Since 2006 the Australian Bureau of Statistics (ABS) and the Department of Immigration and Border Protection (DIBP) have defined NOM to include all persons entering and leaving Australia regardless of their visa status, as long as they

Table 4.1: Components of population change, Australia

Year	000s			Per cent	
	Natural increase	Net overseas migration	Total increase	NOM growth rate	Total growth rate
2008	155.8	315.7	459.5	1.50	2.19
2009	159.2	246.9	390.0	1.15	1.82
2010	158.0	172.0	306.8	0.79	1.40
2011	156.1	205.7	347.8	0.93	1.57
2012	161.8	241.2	402.9	1.05	1.79

Source: ABS, *Australian Demographic Statistics*, Catalogue no. 3101.0, December 2013.

meet certain criteria regarding their length of stay in or absence from Australia. To be included as an addition to Australia's resident population, a person (whether Australian- or overseas-born) has to have stayed in Australia for 12 months out of the 16 months following their arrival in Australia. Conversely to be subtracted from the resident population a person (again whether Australian- or overseas-born) must have left Australia for at least 12 months out of the 16 months after leaving. Thus an Australian-born person or permanent-resident migrant who had previously left Australia and had met the criteria for subtraction from the resident population would be counted as an inclusion if he or she returns for the requisite period.

So NOM measures the movement of everyone coming and going from Australia. For years there have been worries that Australia was losing many of its tertiary-trained residents – the so-called 'brain drain'. Since 2006 the Australian Bureau of Statistics (ABS) has been able to accurately measure the actual scale of such losses. It does so through DIBP's designation of a unique identifier on the visa or passport. This enables the organisation to track the actual length of stay or departure of each mover. If the NOM numbers look high, it is not because they are estimates. They are accurate counts.

Previously, DIBP relied on the statements of movers' intentions to stay or leave when they filled in their departure or arrival cards. When checked against actual movements these statements proved to have exaggerated the numbers leaving permanently and thus the scale of the 'brain drain' (Birrell and Healy 2010). Worries on this count have even less foundation currently given the increasing gulf (documented below) between Australia's attractiveness as a destination relative to other developed countries.

The extraordinarily high level of NOM in the mid-2000s when it reached 315 700 in 2008 (Table 4.1) and again when NOM surged from 172 000 in 2010 to 241 200 in 2012 was mainly due to movements of migrants holding temporary visas. These movements, as we will see, are a product of the policy settings put in place by successive Coalition and Labor Governments and the keenness of prospective migrants to take advantage of the opportunities offered.

If NOM does remain around the 241 200 level reached in 2012 through to 2050, Australia's population will grow from 23 million in 2012 to well over 40 million by 2050. This is way above the controversial 'Big Australia' outlook of around 36 million that was first put on the public agenda in 2010 with the publication of the *Third Intergenerational Report* (Treasury 2010). This outlook was based on the assumption that NOM would be maintained at 180 000 per year.

The fact that current levels of NOM, though way above what was deemed controversial in 2010, have attracted little current public debate is an indication of how far Australian opinion makers have moved in embracing or tolerating such big numbers since 2010.

For its part, DIBP is forecasting that, with the policy settings in place as of mid-2013, NOM will reach 250 200 by 2016 (Table 4.2). This forecast is based on DIBP's expectations for movements in and out of Australia on the assumption that there will be no changes in the rules governing Australia's various permanent- and temporary-entry visas over the years to 2016.

DIBP is also assuming there will not be any sharp changes to Australia's economic setting. A crucial feature of this setting is the extraordinary surge in Australia's prosperity in recent years as a consequence of the mineral investment boom since 2003. This has led to a growing gulf between wage levels and employment opportunities in Australia, not just with the major developing countries of Asia from which most of Australia's migrants are drawn, but also with most developed countries. This situation has increased the attractiveness of locating in Australia permanently or temporarily in order to access our labour market on the part of prospective migrants from both sets of countries. It has also further diminished any concerns about keeping residents (whether Australian-born or migrants) in Australia.

One measure of this attractiveness is the World Bank's measure of Gross National Income (GNI) per capita – expressed in US dollars. On this metric per capita GNI in Australia grew from $41 980 in 2008 to $59 570 in 2012, by which time it was well ahead of most Western European nations. These include the United Kingdom ($38 250) Italy ($33 840) Ireland ($38 970) and even the United States ($50 120).

Table 4.2: DIBP forecasts of NOM to June 2016

	Year to June	000s
Forecast	2013	231.3
Forecast	2014	241.7
Forecast	2015	249.9
Forecast	2016	250.2

Source: Department of Immigration and Border Protection (DIBP), *The Outlook for Overseas Migration*, June 2013.

Sources of NOM

The sources of the surge in NOM can be tracked to both the permanent and temporary visa programs currently in place. These could change; especially should Australia's current relatively buoyant economy subside. However, the willingness of successive governments to create these migration opportunities is a good indicator of policy makers' priorities. So is the willingness to keep these settings in place, even as economic conditions in Australia have softened since late 2011 (a point revisited at the end of this chapter).

The permanent program

Australia's permanent visa program (including the humanitarian program) increased from 184 000 places in 2008–09 to 214 000 in 2012–13. This is higher than at any point in the past three decades. This situation is a product of both the Coalition Government (to November 2007) and subsequent Labor Government's anxiety to meet employer demands for more skilled workers since the beginning of the mineral investment boom in 2003. Since coming to office in late 2013 the Coalition Government has maintained this program target.

The record high permanent program also reflects the demand for entry into Australia. This is obvious in the case of the humanitarian component which Labor intended to increase from an annual figure of around 13 000 in earlier years to 20 000 in 2012–13 in order to accommodate the influx of asylum seekers arriving by boat.

This demand is also evident in the growth of the family reunion component. This may surprise readers since eligibility is largely limited to spouses. There is a parent program, but it is capped at a low level (relative to the demand for parent places). Nevertheless, spouse visa numbers are large and growing. There were 42 098 spouse visas issued in 2007–08 and 50 375 in 2012–13 (DIBP 2013a). Their number will increase as Australia's recently arrived migrant population grows (because of the additional linkages to possible spouses in overseas locations) and because of the rewards delivered (and thus attractiveness of a marriage or de facto marriage offer from an Australian resident) as a result of permanent residence in Australia.

Temporary migrants

As noted, most of the recent growth in NOM is attributable to the influx of overseas-born persons holding temporary visas. The main source of this growth in the 2000s, which culminated in the huge 315 700 estimate for NOM in 2008, was overseas students (Birrell and Healy 2010). Their number surged with the explosion of enrolments in those taking vocational courses, mainly in cooking and hairdressing. The attraction of this qualification was that it provided a cheap entry

Table 4.3: Temporary entrants in Australia as of March 2012 and March 2013

	Mar-2012	Mar-2013	Change
Students	344 480	332 470	−12 010
Visitors	220 380	248 250	27 870
457s	160 420	190 920	30 500
Working Holiday Makers	142 600	170 700	28 100
Bridging visas	132 320	118 820	−13 500
Temporary graduate visa holders	27 980	41 090	13 110
Others	28 670	30 310	1640
Total	1 056 850	1 132 560	75 710
New Zealanders	616 110	632 890	16 780
Total temporaries	1 672 960	1 765 450	92 490

Source: DIBP, *Temporary Entrants in Australia*, 31 March 2013.

point to Australia's labour market, either on a temporary basis while the migrant was enrolled as a student, or as an onshore entry point to a skilled permanent visa, as was possible until this was largely removed with reforms introduced in 2010.

Since the Global Financial Crisis (GFC) in 2008–09, the numbers of temporary visas issued have surged again, this time mainly in the Working Holiday Maker (WHM) category and among those sponsored by employers for temporary work (the 457 visa) and visitors. The number of persons holding these (and some other temporary visas) in Australia (including dependents) as of March 2012 and 2013 are listed in Table 4.3.

The enormous total amount of temporary residents in Australia of 1.7 million in 2013 shown in Table 4.3 includes 632 000 New Zealand citizens. New Zealand citizens can come and work in Australia without a visa and stay indefinitely. Since 2001 they cannot obtain Australian citizenship unless they first apply for permanent residence under one of Australia's permanent visa subclasses. Few have sought to do so. Nonetheless, their numbers continue to grow by 20 000–30 000 per year (offset somewhat by a numerically far smaller flow of Australian citizens to New Zealand). The number of New Zealand citizens is growing, and will continue to grow, as long as the wage gulf and job opportunities in Australia significantly exceed those offered in New Zealand.

One might think that the number of temporary visa holders (other than New Zealand citizens) would be fairly stable because the visas in question are for temporary stays. However, the stock of the major temporary visa subclasses in Australia (with some exceptions) continues to grow, as is indicated by the 92 490 increase between March 2012 and March 2013 shown in Table 4.3. This is partly because the number of temporary visas issued each year is increasing, and partly because the Australian Government has facilitated visa churn, that is, opportunities for persons arriving on a temporary visa to prolong their stay by transferring to another visa.

Again, this outcome is partly a product of deliberate Australian Government policy to liberalise access to temporary visas (detailed below for the WHM and 457 visa subclasses). The huge 1.7 million stock also reflects interest from prospective migrants in eligible source countries to take up the opportunity to access Australia's labour market that these visas provide. The great majority of the current stock of temporaries as of March 2013 were on visas which permitted the visa holder to work (if on a temporary basis) in Australia. An additional inducement (again, detailed below) is that once in Australia, temporary visa holders are in a better position to obtain a permanent resident visa.

Working holiday makers

The WHM program is uncapped except for a tiny sub-program called the Work and Holiday visa subclass. It is therefore demand driven in the sense that the numbers visaed depend on the interest of young people (aged 18–30) from eligible countries to apply. Again, except for the Work and Holiday visa subclass, there are no skill, English-language or education requirements. The eligible countries do not include developing countries such as India and China. However, South Korea, Taiwan and Hong Kong are eligible. So, too, are most Western European countries (though not yet Spain – however, negotiations to include Spain are underway), including those with acute youth unemployment problems, notably Ireland and Italy, where there is an obvious incentive to move – if temporarily – to access job opportunities in Australia.

The WHM program was liberalised in 2005 so that WHM visa holders who work for three months in regional areas (initially in horticulture, later extended to mining and construction) were allowed to extend their stay beyond one year to another year. WHMs can work the entire time they are in Australia in employment, though they are restricted to a limit of six months with one employer.

The number of WHM visas issued has jumped, from 192 922 in 2010–11 to 258 248 in 2012–13. The 2012–13 figure includes about 40 000 who worked in a regional area and thus were eligible to obtain a second one-year visa. As might be expected, there has been rapid growth in the number of young people escaping the jobs crisis in places like Ireland and Italy. However, the most notable growth has been from South Korea and Taiwan. In 2012–13, 35 761 WHM visas were issued to citizens of Taiwan and 35 220 to South Koreans. Their numbers are approaching the 46 131 issued to the once dominant source country of the United Kingdom (DIBP 2013b).

Temporary migrants on 457 visas

It is has long been Australian policy to permit employers to sponsor migrants to work for temporary periods, but only if in doing so the interests of domestic workers were protected. However, in 1996 the rules governing employer sponsorship for temporary entry work visas were deregulated. (The new rules were proposed by the Keating Labor Government and implemented by the incoming 1996 Coalition Government.) Previously, employers had to provide evidence that they had tested the Australian labour market before sponsoring a migrant on a 457 visa. The new rules abolished this requirement. Employers were henceforth allowed to recruit whoever they pleased, effectively in as many numbers as they wished as long as the jobs in question were in occupations requiring trade level skills or above.

The context in which this deregulation occurred was the commitment of the Hawke/Keating Labor Governments to open up the Australian economy to global competition. The prevailing view within policy-making circles was that, as part of this opening up, Australian-based enterprises should be able to draw on the skills they needed from wherever they chose. This stance remains. It was reflected in the Coalition Government's decision in 2005 to offer to eliminate Australia's right to labour-market test temporary workers during the Doha round of international trade negotiations, in return for trade concessions. This bargaining offer was affirmed by the Rudd Labor Government in 2008.

The number of 457 visas issued to principal applicants grew strongly at the end of the 2000s from 48 080 in 2010–11 to around 68 000 in both 2011–12 and 2012–13. This surge occurred despite the softening of the Australian labour market after the peaking of the mineral investment boom late in 2011. An analysis of why this occurred illustrates some of the dynamics of Australia's migration policy settings, in particular their built-in potential to attract ever-larger numbers of temporary migrants.

An important background policy setting is that since the mid-2000s it has been Australian Government policy to outsource an increased share of the skilled migrants under the permanent entry program to employers. The rationale was that employers were the best judges of the migrant skills needed. Migrants sponsored by employers now make up the largest component of the permanent entry skilled program. Successive policy changes since the mid-2000s have also encouraged employers to sponsor temporary migrants already in Australia (Birrell *et al.* 2011, p. 34).

As we have seen, the stock of temporaries in Australia is high and many are interested in obtaining permanent residence (as through sponsorship by an employer). From the employer's perspective, the benefits of sponsoring an onshore migrant include avoiding the costs of recruiting overseas and providing transport for the sponsored migrant to Australia.

The current rules also stipulate that, to be eligible for employer sponsorship for permanent entry, a 457 visa holder must have worked for the employer for at least two years. This advantages the employer because, if the 457 visa holder wants to be sponsored, they will be under pressure to accept the terms and conditions of employment offered while employed by the sponsor. For employers operating in highly competitive labour markets, like construction and the hospitality industries, sponsorship and retention of 457 visa holders offered a competitive advantage – thus helping to explain the continued rise in 457 visa recruitment, despite the softening of Australia's labour market since late 2011.

By 2012–13 some 53 per cent of all 457 visas sponsored were already in Australia on temporary visas when sponsored, up from 43 per cent in 2011–12 (CFMEU 2013, pp. 6–7). In turn, the great majority of those subsequently sponsored for permanent entry by employers were already in Australia and most were already working for the employer on a 457 visa.

Successive Australian Governments had created a recruitment circle, starting with policies facilitating high temporary migration that then fed into permanent migration, which thereby encouraged further temporary migration because of the pathways to permanent residence.

These outcomes were not unintended consequences of separate policy decisions. The idea that temporary migration for work would be genuinely temporary, a stopgap while domestic workers were trained, was jettisoned with the 1996 changes. The attitude amongst policy elites now is that Australia should attract skilled migrants whatever way it can and then keep them.

These arrangements have been resisted by Australia's trade unions. They have insisted that the 457 visa program and the other temporary entry programs (including WHMs) have harmed domestic-worker opportunities and that such workers should have a legislated right to first access to available work.

In 2009, as a response to union pressure, the Labor Government did increase the minimum levels of wages that employers must pay a 457 visa holder. It also raised the minimum English-language level required. Just prior to the September 2013 election, Labor also legislated to require labour market testing for at least some of the applicants for 457 visas.

On the other hand the Labor Government resisted pressure to put a cap on the number of 457 visas issued and refused to introduce a skills occupation list which actually took account of whether the occupation was oversupplied in Australia or not. This, had it been introduced, would have removed occupations like accounting, which by 2012 the Labor Government's own labour market specialists had determined were in oversupply (DEEWR 2012).

For its part, the incoming Coalition Government in 2013 has indicated that it favours employers' rights to sponsor migrants on 457 visas with minimal constraints (including minimal labour market testing). The Coalition has also indicated that it favours using access to the Australian labour market as an instrument to promote the overseas student industry.

The momentum towards higher migration

If the recent political record is any guide, those concerned about the environmental and other costs of migration have little influence on Government policy. The aftermath of the debate about a 'Big Australia' is a case in point. During the 2010 election campaign the new Prime Minister, Julia Gillard, announced that she did not favour a 'Big Australia' largely because of voter concerns about congestion and escalating housing prices in Australia's metropolises. An inquiry into *A sustainable population strategy for Australia* was commissioned which produced an issues paper with several appendices (Commonwealth of Australia 2010) but little action, except to channel a slightly larger number of migrants into regional areas. Since that time, as we have seen, the temporary migrant influx has mushroomed and the permanent entry program has increased from 184 000 in 2008–09 to 214 000 in 2012–13.

Australia has experienced a sharp slowdown in net job growth since late 2011. This has meant increasing difficulties for domestic job seekers. But these difficulties have not resulted in any slowdown in Australia's record high migration, nor did they result in any serious political debate during the 2013 election (aside from that relating to the 457 issue described above).

It might be expected that Labor Party leaders would have exploited the issue in order to secure the party's working-class constituency during the election, given that the Coalition was taking a strong stand on maintaining high migration.

That they did not do this is at least in part a consequence of the ideological commitment of most Labor leaders to an Australian future marked by cultural diversity, openness to the global marketplace and to population growth. To make immigration an issue is, according to this perspective, to risk taking a backward step on all these fronts.

After enduring poor results in successive opinion polls for much of the period since 2010, as well a disastrous electoral thrashing in the 2013 election, the party leaders are in search of new mission. They have found it in a revival of what is now regarded as the Party's greatest era of reform. This was the Hawke/Keating era of government from 1983 to 1996 when Hawke and particularly Keating led a revolutionary rejection of Australia's protectionist, defensive stance to the world. This period saw the institutionalisation of a series of reforms (including the float of the currency, the encouragement of foreign investment and the slashing of tariffs).

Looking back on this era in 1996 (after a Labor electoral defeat on the scale of that delivered in 2013), Paul Keating laid down how he saw migration policy in the context of his legacy in globalising the Australian economy. According to Keating, Australia had only one viable future; that was as an exporter into Asia. To do this successfully must involve an embrace of migration and multiculturalism. Keating saw this as a symbolic commitment, central to sustaining the globalisation impetus. To go backwards would be to retreat into our previous 'third rate' condition (Keating 2011, p. 144). He declared that 'Australia's post war immigration policy was one of the greatest strategic decisions this country has made' (Keating 2011, p. 145). He asserted that we must press ahead with this policy (notwithstanding the support Pauline Hanson was attracting at the time) because he saw it as inseparable from Australia's success in the global economy.

This is the way contemporary Labor elites see immigration. Most, including the new leader Bill Shorten, see a commitment to high immigration as part of the Keating heritage that they believe will restore their claim to be taken seriously as innovative reform leaders. The former high profile ABC current affairs anchor and vanquisher of John Howard in the 2010 election, Maxine McKew, provides a telling illustration of this sentiment. Not surprisingly, as a Rudd supporter, she is hostile to his successor, Julia Gillard. Gillard's worst sin is said to be her criticism of a 'Big Australia' (McKew 2013, p. 19). By mounting this criticism, so McKew explains, Gillard betrayed the Keating legacy and, in the process, Labor's strongest claim to leadership in Australia.

While Labor's leaders embrace this message they will not challenge Australia's current population outlook as described above.

Nor will Australia's business leaders. Australia's role in the global economy has so far been that of a supplier of commodities. Rapid population growth does not seem consistent with an economy specialising in capital-intensive yet low employment industries. Rapid population growth implies the dissipation of the wealth generated to accommodating the increased population. Yet Australian business leaders are as one about the need for high migration. In the present context, where the mineral investment boom has crested, they see the revival of housing and city building as the answer. For this to occur, they argue, migration must continue.

This is a commitment written deep into the heart of Australian capitalism, most centrally within the four major banks that now dominate the economy and within opinion in elite business circles including the Business Council of Australia. The banks' extraordinary escalation in profits depends on growth in their major market, which is mortgages on residential property.

This is not to say that tensions over employment, urban quality of life and natural environment issues may not provide a rallying point to contest the left and right commitment to a 'Big Australia'. It is to say that it will not be easy to rally opposition, especially while Australia's high per capita wealth and relatively low unemployment rates continue. If such opposition does occur it will be in the face of media disapproval. In this context perhaps the last word should go to the media baron who dominates the print media in Australia. Rupert Murdoch stated in a speech to the Lowy Institute on 31 October 2013:

> *The nations that lead this century will be the ones most successful at*
> *attracting and keeping talent. There are countless thousands of intelligent*
> *university graduates around the world, and in particular in our Asian*

neighbours, looking for work, and wanting to start businesses. We need to get the brightest of them here. (Murdoch 2013, p. 6)

References

Birrell B, Healy E (2010) Net Overseas Migration (NOM): why is it so high? *People and Place* **18**(2), 63–64.

Birrell B, Healy E, Betts K, Smith TF (2011) *Immigration and the Resources Boom Mark 2*. Centre for Population and Urban Research, Monash University.

CFMEU (2013) *CFMEU Analysis of 457 Visa trends, Report No 1*, CFMEU National <http://www.cfmeu.asn.au/sites/cfmeu.asn.au/files/downloads/nat/campaign/cfmeu-457-analysis-30-october-2013.pdf>

Commonwealth of Australia (2010) *A Sustainable Population Strategy for Australia*. Department of Sustainability, Environment, Water, Population and Communities, Canberra.

DEEWR (2012) *Labour Market Research – Accountants, 2011–12*. Department of Education, Employment and Workplace Relations, Canberra

DIBP (2013a) *Annual Report, 2012–13*. (The 2012–13 partner total includes 4000 visas granted as a result of the recommendations of the Expert Panel on Asylum Seekers in 2013.) Department of Immigration and Border Protection, Canberra.

DIBP (2013b) *Working Holiday Maker Visa program report, 30 June 2013*. Department of Immigration and Border Protection, Canberra.

Keating P (2011) *After Words*. Allen & Unwin, Sydney.

McKew M (2013) *Tales from the Political Trenches*. Melbourne University Press, Melbourne.

Murdoch R (2013) *2013 Annual Lowy Lecture, Tenth Anniversary*. Lowy Institute, Sydney.

Treasury (2010) *Australia to 2050: Future Challenges (Third Intergenerational Report)*. Department of Treasury, Canberra.

5

What population growth will do to Australia's society and economy

Mark O'Connor

Growth is the problem to which it pretends to be the solution. (William Grey, University of Queensland)

Australians know their country has a growing population, yet the media rarely mention how extreme our growth rate is. The Australian Bureau of Statistics (ABS) says our annual population growth has lately averaged around 1.8 per cent. That's more than four times the average of industrialised countries, and high even by Third World standards. For example, Indonesia's population growth of 1.1 per cent per year is far lower – yet it sees its rate as too high and seeks to reduce it (Hitipeuw 2011). So we are out on a limb. Our growth is pushed along by what is close to the world's highest per capita immigration program[1] and until recently by 'baby bonuses'.

Why do we do it? Some people imagine that Australia could solve the rest of the world's environmental problems by taking more immigrants. This is not so. India and China together grow each year by about the total population of Australia.

Let us also remember that the overpopulation of Asia, though gross, is quite recent. Java, which held 124 million humans in 2005, held less than five million in 1800. This growth was produced, in two centuries, by an average annual increase less than Australia's recent rates (O'Connor and Lines 2010). The human history of even the most crowded countries starts with a small population and usually with an intact environment and abundant natural resources. Yet once countries pass 20 million, as Australia did in 2003, they are out of the tiddlers' pool – and only two or three doublings away from becoming another obese behemoth of around 100 million.

Studies of how civilisations fall – such as *Collapse*, by the American scientist Jared Diamond (2006), or Joseph Tainter's *Collapse of Complex Societies* (1988), or Claire Russell and WMS Russell's *Population Crises and Population Cycles* (1999) – show a common cause which is that populations outgrow their resources. The collapse tends to occur just when expansion has been most successful. That is, civilisations do not so much decline and fall as rise and topple – just when they have never seemed bigger or better. The empire of 100 000 persons in a fertile valley is vulnerable to a once-in-200-years drought; but the empire of 200 000 persons in the same valley may collapse from a mere once-in-20-years drought.

Yet for many or most economists, annual population growth at the typically Third World rates experienced recently by Australia implies prosperity – via a steady growth of GDP. For mathematicians and statisticians, like Albert Bartlett, such rates imply exponential growth – impossible on a finite planet.

For neo-liberal theorists these same rates of growth imply a first law of anti-Malthusian experientialism: that exponential population growth has so far done us much less harm than good, and always will, because human ingenuity always prevails. For biologists it demonstrates rather the principle of human selfishness: that humans will chop down the Amazon rainforest and exterminate a cascade of other species sooner than see the price of soybeans, and therefore of bacon, go up. In the short run this principle of human selfishness triumphs. But in the long run Bartlett's exponential principle will also win. Stabilising the world's and our nation's population now may be too late to save us and other species in the long term, but it would surely assist us to:

- relieve overstretched infrastructure
- ease cost of living pressures
- protect our environment, and limit carbon emissions
- promote education, training, and employment of young adults
- minimise overdevelopment, whether high rise or sprawl, and
- create a more resilient economy that does not depend on a resources boom.

That constant growth on a finite planet is a mathematical impossibility was well brought out in an exchange between the late Julian Simon, an economist who believed in indefinite growth, and Bartlett.

As Bartlett puts it, 'Humanity's greatest failure is its inability to understand the exponential function.' Take a case in point. If Melbourne is currently four million people and continues at 1.8 per cent p.a., it would be 5.4 million by 2030, 7.7 million by 2050, and 18.9 million by 2100 – and so on, ever higher.

Yet we have Australian exceptionalism and endless growthism written even into our national anthem, with the assurance that we have 'boundless plains to share'. We sing this, despite the fact that all our farmland was essentially settled over a century ago, and that recent arrivals almost never become farmers.

So what are we in for over the next 80 years or so?

Well, Australia will suffer more than most countries from climate change, and may even lose the ability to feed its projected future population of some 40 million by mid-century. I am not sure, however, that we will see any very clear effects upon the world's or Australia's climate from our inaction on our own emissions. Climate change is, in that sense, a global problem.

Yet we are already feeling, quite close to home, the many effects of our Third World rate of population growth. The resulting social effects, for most of us, might start with some mention of the environmental damages.

As the UK's Chief Scientist David King (2006) remarked:

It is self-evident that the massive growth in the human population through the twentieth century has had more impact on biodiversity than any other single factor.

Each extra human adds to the pressures on other species. *Yes, kangaroos and koalas, every human is doing you damage. Every extra human.* As Queensland's former Minister for Sustainability, Andrew McNamara, puts it: 'Any talk of sustainability without a commitment to population stabilisation is not just spin; it is a dangerous lie.'

More people mean more roads, houses, offices, hospitals and shopping malls, and that means we clear more native vegetation (a prime cause of extinctions) on what is often the richest land or the coastal fringe. Greenhouse gas emissions also rise (Sobels *et al.* 2010). Prime farmland becomes housing estates.

Hence in 2010 Australia's largest conservation body, the Australian Conservation Foundation, formally requested the Federal Minister for Environment to classify Australia's population growth as a threatening process for the environment.

Let us turn now to the purely human implications of continued population growth on this semi-desert continent.

Food and water and energy

A recent report on the long-term implications of immigration, by the National Institute of Labour Studies at Flinders University and the CSIRO's Sustainable Ecosystems, warns: 'The security of production of food in Australia (and imported from overseas) is in question.' The report, prepared for the Department of Immigration, sees our oil situation as grim (Sobels *et al.* 2010). If we run net migration at 260 000 per year, then by 2050 we will need to import about twice as much oil as now. Other likely changes include 'a doubling to a tripling of greenhouse emissions ... increased traffic congestion and critical water shortages in three capital cities' (Foran 2011). Australia's population of 23 million is heading for some 36 million by mid-century. Yet former treasury Secretary Ken Henry estimates that a sustainable population for Australia, with business as usual, would be 'about 15 million' (Henry 2011).

The global economy now has its head in a sort of 'oil noose'. If the economy grows, it pushes up the price of oil, which tends to choke economic growth. The basic assumption of growth economics, that growth can go on forever, begins to seem like folly. Capitalism seemed a robust system while it could grow and feed on a seemingly limitless natural world. Yet now we are facing peak oil, peak gas, peak water, peak fish, peak phosphorus – the list goes on (Heinberg 2007; Déry and Anderson 2007).

As for Australia's situation, Treasury's third *Intergenerational Report* contains a simple stunning statement, followed by an equally stunning silence: Australia's oil will be gone by 2020 (Department of Treasury 2010). And we are fast selling off our vast reserves of gas.

Security

In 2010, the former Chief of the Australian Defence Force, Admiral Chris Barrie, warned a Davos conference of Australian business and government leaders of 'three elephants in the room': population, the economy (neither can just keep on growing) and climate change.[2]

Our cities suck in energy and food. Until recently, both have been absurdly cheap. We have used fossil fuels like a lottery winner spends cash. New suburbs have been recklessly multiplied on the assumption that oil and food could easily be brought to them.

Yet Australia's cities now have few factories. Overall, they 'export' too little economic product (unless you count booming house sales to newcomers) to pay for the necessities they must import. They have followed the Las Vegas model as if economic growth, based on property speculation and house-building, could somehow support people even in a desert. Melbourne is particularly vulnerable (Birrell and Healy 2010).

Peak Oil makes it crucial to ask of each new proposed suburb: What commodity will the residents be so exceptionally well-placed to produce and export that the rest of the world (or even the rest of Australia) will be disposed to send them treasures like oil and food in exchange? We must also abandon the foolish notion that making a city denser by adding more people will make it more 'sustainable'.

Infrastructure and services

Population growth also has a huge economic effect. As Jane O'Sullivan has pointed out,[3] our rapid growth runs up an impossible infrastructure bill. Each extra Australian, whether born or migrating here – and regardless of whether they move here from a rich or a poor country – requires at least some $200 000 worth of additional infrastructure. According to Infrastructure Australia, which advises the Federal Government on infrastructure projects, Australia already has a $770 billion backlog in public and private infrastructure (Hepworth 2010). Even the tax revenue provided by the mining boom (Bartlett 2007) can only partly conceal the impoverishment due to population growth. Basically we have spent the profit of the mining boom on expanding our population and thus our vulnerability. We might have done better to have pissed it up against a wall.

Councils and city planners now find that, as the Red Queen told Alice in *Through the Looking Glass*, you have to run as fast as you can simply to stand still – or not to go backwards. Governments are so short of money that desperation often trumps planning; they sell off public property, or fail to conserve crucial areas, just to balance the books. Quality of life falls, and voters react.

A principle you may discern here, is that for other issues, such as the survival of other species or the availability of resources, the pinch comes from the absolute size of the population. But for infrastructure, the most crucial thing may be the *rate* of population growth. As Kelvin Thomson argues:

> *A society with a stable population needs to replace 2 per cent of all infrastructure annually. But if a population is growing at 1 per cent per annum, for example … this increases the burden of infrastructure creation by some 50 per cent. … 1 per cent more GDP or tax cannot pay for 25 to 50 per cent more public infrastructure. (House of Representatives, 19 March 2012)*

True, a bigger population means a bigger gross domestic product (GDP) but it doesn't mean more GDP per person. And measurements of GDP don't take into account so-called 'externalities' – such as free access to amenities like sports grounds and bushland – which come under increased pressure as population grows. Amenity and environment are constantly eroded in the name of economic growth.

Economic rocks ahead

In addition to the cost of community infrastructure, at some $200 000 per extra person, each new citizen becomes part owner of Australia's finite mineral wealth, which not only supports our current standard of living, but is needed to fund the transition to a post-minerals-boom economy. (Translation: to keep our children in the First World.)

In effect, each new citizen gets a shared inheritance worth a vast sum. This is not money in the hand, but it is still real wealth, and a doubling of population effectively halves the existing population's share. Amazingly, Treasury provides no estimates for either of these two huge costs.

It is clear that the costs dwarf the gains – except to employers who get cheaper labour, property investors who get increased land values, and banks which reap more or larger mortgages. If employers had to repay the real costs to the public purse, they would indignantly reject the deal. The same applies to colleges and universities that profit from students who are, in effect, buying permanent residency by doing a course. About $2000 profit per student per year may be costing the rest of us hundreds of thousands.

House prices

Another effect of such rapid population growth is that house prices cripple many families. Marriages crack under the strain as people feel their lives slipping away from them, while working to pay off a house in a nice suburb. Or any suburb.

I lived for three years in a part of Tuscany where population was stable. I was struck by the fact that none of the couples I knew there seemed worried about buying a house. There was always a spare house somewhere in the family. In Australia today you expect to have to buy a house. In a stable population you expect to inherit one.

Back in Australia, population pressure and urban densification produce ever-worsening traffic jams that merely add to the time parents spend away from home.

Because wages have to cover mortgages, absurd house prices add to the cost of every product and service we buy. Also, investment gets diverted from producing goods and services to property. True, your house may wind up worth millions; but while you still need to live there, you can't sell it. Most baby boomers, it seems, plan to 'age in place'.

Property cost is an invisible impost that impoverishes us all. It keeps us working – and driving – for ever-longer hours, without seeming to get the quality of life our parents' generation had. In late March 2011, *The Age* reported that,

'Australia is in a housing affordability crisis, with even the last bastion of cheap housing – new suburbs on city fringes – moving beyond the reach of most first home buyers.'

Yet the Treasurer Joe Hockey confirmed in October 2013 that he favours high immigration as a way of continuing to push up house prices (AAP 2013)!

Sydney now ranks second only to Hong Kong as the most unaffordable housing market in the English-speaking world, with Melbourne just three places behind. The high immigration, demanded by employers wanting to hold down the price of labour, reduces wages and also escalates house prices, creating a mortgage trap that spreads stress and depression, mangling families. It also makes us a less equal society. As former *Canberra Times* editor Crispin Hull puts it: 'The population growth scam … enriches and aggrandises the few, and impoverishes the many.'

Politicians pretending not to see

Get thee glass eyes
And, like a scurvy politician, seem
To see the things thou dost not.

King Lear Act 4, Scene 6.

– or seem not to see what they do see.

One political leader who seems to grasp this problem is Labor's John Robertson, currently the NSW Opposition Leader. In a recent article in the Sydney *Telegraph* titled 'Sydney needs services, not just a rising population', Robertson gets as far as saying that 'The central policy challenge for NSW is maintaining our quality of life in the face of a rapidly rising population.' As he points out:

Everyone knows Sydney is already bursting at the seams. Hospital waiting
rooms are full. School class sizes are on the up. Our suburbs are inching
further out. Yet all of this is before our State braces to absorb another
2 million people between now and 2031 … the Government plans to cram in
another 43 000 people [in Sutherland Shire]. Yet it has no concern for how
this will affect quality of life in one of the most precious and unique areas of
Sydney.' (Robertson 2013)

This puts Robertson way ahead of two of Labor's federal leaders, Bill Shorten and Paul Howes, who are still singing the praises of population growth.

Of course John Robertson did not suggest that he would do anything concrete to stop this rapid population growth. If he had done so, Murdoch's editors would almost certainly have refused to run his piece.

I cannot think of a better example of the quagmire of stymied thinking in which Australia now stagnates. Our children and grandchildren will pay a heavy price for this.

Endnotes

1   But see The Population Reference Bureau's 2010 world data sheet. It gives Australia's immigration rate as 13 per 1000. A few countries are higher: Bahrain 42, Oman 26, Singapore 25, Luxembourg 16, and Guam is 13. However, migrants to Gulf States may not be migrants in the sense of having the right to remain when their work-contracts end. Migration to Singapore has not been consistent. The average for the 'more developed countries' is 2 per thousand. See <http://www.prb.org/Publications/Datasheets/2010/2010wpds.aspx>.
2   Author's notes, confirmed by Admiral Barrie's email of 6 May 2011.
3   See: <http://www.pc.gov.au/__data/assets/pdf_file/0004/135517/subdr156-infrastructure.pdf>

References

AAP (2013) Hockey prepared for US default. *Business Spectator.* <http://www.businessspectator.com.au/news/2013/10/15/politics/hockey-prepared-us-default>.

Bartlett A (2007) A depletion protocol for non-renewable natural resources: Australia as an example. *Natural Resources Research* **15**(3), 151–164.

Birrell B, Healy E (2010) Melbourne: a parasite city? *People and Place* **18**(2), 52–67.

Department of Treasury (2010) *Australia to 2050: Future Challenges (Third Intergenerational Report)*, p. 91, Canberra.

Déry P, Anderson B (2007) Peak phosphorus. *Energy Bulletin*, 13 August.

Foran B (2011) Population policy is driven by the Dolly Parton syndrome. *The Punch*, 9 March 2011.<http://www.thepunch.com.au/articles/australias-population-policy-is-like-dolly-parton/>.

Heinberg R (2007) *Peak Everything: Waking Up to the Century of Declines*. New Society Publishers, Gabriola Island, BC, Canada.

Henry K (2011) Lateline: Henry takes parting shot at complacency (transcript). <http://www.abc.net.au/lateline/content/2011/s3155914.htm>.

Hepworth A (2010) Infrastructure Australia unveils wish list. *The Australian*, 7 June 2010, pp. 21–22.

Hitipeuw J (2011) Indonesia to cut population growth to below 1 percent. *Kompas.com*, 25 January 2011. <http://english.kompas.com/read/2011/01/25/0719093/Indonesia.to.Cut.Population.Growth.to.Below.1.Percent>.

King D (2006) Statement to a British Parliamentary enquiry, 3 July. <http://population-matters.org/2007/press/leaders-urged-brave-population-growth/>

O'Connor M, Lines W (2010) *Overloading Australia: How Governments and Media Dither and Deny on Population*. 4th edition, Envirobook, Sydney.

Robertson J (2013) Sydney needs services, not just a rising population. *The Daily Telegraph*, 3 October 2013, p. 27.

Sobels J, Richardson S, Turner G, Maude A, Tan Y, Beer A, We Z (2010) *Research into the Long-Term Physical Implications of Net Overseas Migration: Australia in 2050.* pp. 109, 111–113. Department of Immigration and Citizenship, Canberra.

Trounson A, Hare J (2011) Foreign fees keeping unis afloat. *The Australian*, 1 July 2011, p. 7.

6

Ageing paranoia, its fictional basis and all too real costs

Jane O'Sullivan

Over the past decade, demographic ageing has become a preoccupation of governments and social scientists globally. It is presented as a threat to prosperity, requiring bold policy measures to moderate and mitigate its impacts. A common response is to boost population growth, through encouragement of larger families and increased immigration quotas. Even among nations whose populations are still growing strongly, and who currently have a small proportion of people over 65 years of age, the fear of ageing has discouraged action to reduce population growth.

At the same time, concerns relating to planetary limits, including food security, water scarcity, loss of natural environments and biodiversity, greenhouse gas emissions and fossil fuel dependence, are becoming ever more acute. Population pressure is the acknowledged driver of all these challenges, but the future projected growth is taken as a fact over which we have no influence. Equally accepted is that this growth will be limited. Most reports refer to nine billion as the maximum number to be accommodated, although this estimate is long out of date.

These contrasting agenda reveal a glaring inconsistency. Population growth is readily accepted as a policy choice, when arguments are made for stimulating it. It is presented as inescapable fate when dealing with problems that would be lessened by reducing it.

The importance of our population choice

It is evident that global population will not stabilise unless individual nations choose to embrace population stabilisation or decline. Their recent actions to prevent demographic 'stagnation' have had global effect. Funding and political support for family planning programs has dwindled while birth rates in many developed countries and some developing countries have risen, with government encouragement. The combined effect has seen global fertility reduction stall, the annual increase in global population creep upward since 2000 and the United Nations' projections repeatedly revised upward. The current medium projection would climb beyond 11 billion early next century.

Yet the conditions required for the medium projection are still not met. Announcing the new UN projections, a refreshingly direct John Wilmoth, head of the UN Population Division, stressed that the medium fertility projection assumed steady fertility decline, and 'is thus an expression of what *should be possible* …[It] could require additional substantial efforts to *make it possible*' (Wilmoth 2013, emphasis in the original).

Suffice to say that, without 'additional substantial efforts', the global population is on course for well over 11 billion. Few analysts of food security consider that we are likely to be able to feed such a number. The more likely outcome is that planetary limits will cause the death rate to climb.

The population projections are blind to such limits, as they only project past trends which have seen life expectancy steadily rise. By choosing to treat these projections as fact, rather than treating future population as a choice, we are likely choosing widespread death by famine or conflict triggered by food prices (as in Egypt), water scarcity (as in Syria) or mass movement of people (as in Central African Republic).

Nations fear ageing more than overpopulation

It would appear from the UN's regular surveys of national population policy that concern about demographic ageing is the primary excuse for revised attitudes to population growth (Figure 6.1) (United Nations Population Division 2011).

It is particularly concerning to see developing countries express concern about ageing, despite these nations having no distinct retirement age and strong workforce participation of older citizens (O'Neill *et al.* 2010). A Myanmar NGO recently claimed 'by 2050 there will be a serious ageing problem in Myanmar with no economically active group to run the country'. Such hyperboles are regularly reported with no attempt at verification. None balance the ageing challenge with

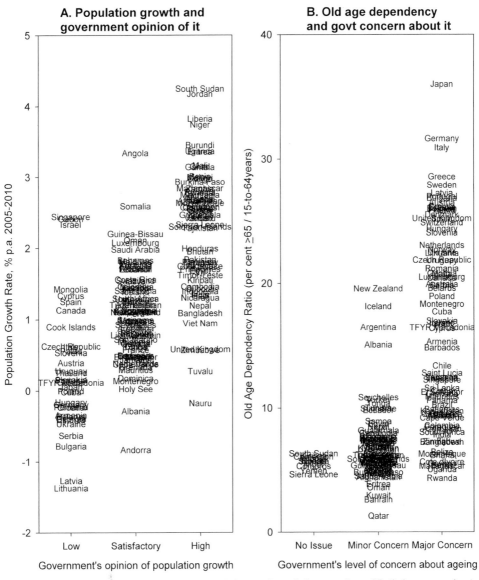

Figure 6.1: Views of national governments on: **A.** their rate of population growth, and **B.** their concern about demographic ageing, compared with their actual population growth and old-age dependency. Overlapping country names are not intended to be legible, but to depict distribution of views. (From United Nations Population Division 2011.)

gratitude for the economic stimulus and food security their rapid population stabilisation enables, nor herald the peaking of their population with enthusiasm.

It is still possible to achieve a peak global population under nine billion, if remaining high-fertility countries follow the example set by strong family planning nations like Thailand, Iran or the Maldives, and if low-fertility countries end pro-natalist policies and embrace their population peak and decline. The advantages of this course for global security are clear. The remainder of this chapter will examine the claimed costs, in terms of more rapid demographic ageing.

Ageing is an inevitable but self-limiting feature of the demographic transition

The demographic transition, from high death rates and birth rates to low death rates and birth rates, is the hallmark achievement of the modern era. Better quality of life both generates it and is enabled by it. Completion of the transition to restore population stability is an absolute requirement of sustainable development. Prolonging an intermediate state, with low death rates but high birth rates, not only leads to an unsustainable population in the longer term, but in the present imposes a burden of population growth *rate* which severely hampers economic development (O'Sullivan 2012).

A direct consequence of demographic transition is a change in the ratios of people of different age. As more people live to an older age, a higher proportion of the population will be old. As people have fewer children, a smaller proportion of the total population will be children. The typical charts used by demographers, which stack the age cohorts vertically with the youngest at the bottom, move from a pyramid shape to a column with a tapered top. If the birth rate is below replacement or significant adult immigration occurs, the base may be narrower than the mid sections, forming a 'coffin' shape. The negative connotations of the word 'coffin' have been used to present this demographic profile as something to be feared. This fear is baseless.

The proportion of people over 65 remains small in the early phases of the demographic transition, as it is young people who benefit most from the initial mortality reduction. If and when the birth rate drops, the proportion in the middle years ('working age') swells. This window of time, in which an abnormally large proportion of people are of working age, is referred to as the 'demographic dividend' of reduced fertility, as it may stimulate economic development if those extra adults are productively employed. Inevitably, however, the increase in the proportion of people over 65 begins to exceed the decrease in children, so that the proportion of working age declines again.

Most developed countries are well into this phase. Australia, USA and Canada, due to sustained population growth, are only beginning to leave the nadir of dependency.

But leave it they will, regardless of future population growth rates. Figure 6.2 depicts the change in old-age dependency and proportion of working age for Australia since 1960, and to 2050 using two projections. One assumes high immigration and population growth, similar to that assumed in the 2010 Intergenerational Report (Australian Treasury 2010). The second assumes no net immigration, generating a total population similar to that assumed in the first

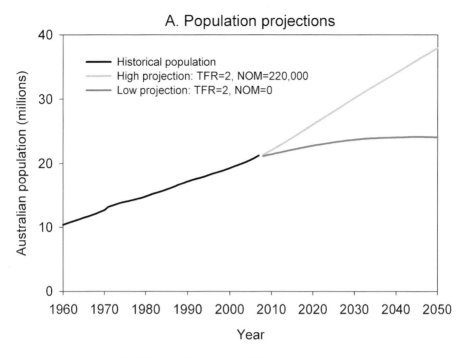

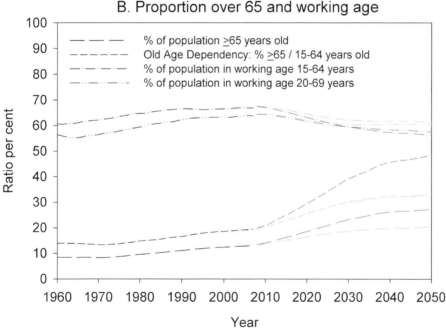

Figure 6.2: A. Australia's population since 1960 (black), and projected to 2050 under either high growth (light grey) or a stabilising scenario (dark grey). TFR = total fertility rate, NOM = net overseas migration. **B**. For each series, the proportion of people over 65, old-age dependency ratio and proportion of working age, defined both as 15–64 years and 20–69 years. (Data from Australian Bureau of Statistics, Population Projections – Australia 2006–2100, Catalogue No. 3222.0, series 5 and 59.)

Intergenerational Report (Australian Treasury 2002), although the latter combined some immigration with a lower fertility rate.

Figure 6.2 illustrates how the conventional measure of 'old-age dependency' exaggerates ageing. Instead of dividing the number over 65 by the whole population, it is divided by the smaller number of working age people. Meanwhile, the proportion of children declines, lessening the change in proportion of working age. Although 'working age' is traditionally defined as 15–64 years, the UN and developed countries accept that 20–69 better reflects modern reality. Using this standard, the proportion of working age only returns to the level experienced in the 1960s, with little difference between the two projections.

Importantly, the changes do not continue indefinitely, but level off. Depending on longevity gains, we might expect between 25 per cent and 28 per cent of people to be over 65 in a stable, healthy population. If the population were allowed to contract at 1 per cent per annum (achievable with about 1.5 children per woman and no net migration) around a third of people may be over 65, assuming life expectancy under 90 years. Having half the population over 65, although often claimed in the media as imminent, is highly unlikely. This implies a life expectancy of 130 years (twice 65), or an extremely low birth rate without immigration, or a significant exodus of younger people, such as some eastern European communities have recently experienced.

The '3Ps': GDP = Population x Participation x Productivity

Population growth as a remedy for ageing was first given prominence in Australian political discourse following Treasury's first Intergenerational Report in 2002 (Australian Treasury 2002). In the subsequent years, the Howard government implemented strong measures to boost population growth, through a 'baby bonus' payment and substantial increases in immigration quotas and temporary work visas. The second Intergenerational Report in 2007 was the opportunity to sell this strategy. It adopted the catchphrase '3Ps', stating that GDP is a product of population, participation and productivity. Population being the easiest of these factors for governments to influence, it was the focus of the greatest shifts in program settings.

This move was at odds with the Productivity Commission (2006), who warned that high immigration was likely to make the average Australian worse off. While per capita GDP might be marginally increased under their assumptions, the beneficiaries would be employers and the immigrants themselves, not Australian wage earners or retirees who would experience depressed wages and increased living costs. The Productivity Commission acknowledged that its analysis did not include impacts on environmental services and amenity. Also missing from their estimation was the cost of infrastructure to provide for additional people.

The 3Ps are built on a set of very problematic assumptions:

1 *Natural resources don't matter.* Diluting, degrading and depleting them will not affect productivity, wealth or wellbeing, because they are not in the model.

2 *Job seekers create jobs.* The size of the economy will be proportional to the number of working age people. Just add people, and the market will do the rest.

3 *The three factors are independent.* If the formula is interpreted to advocate boosting any one of these factors, it must be assumed that boosting it will not be to the detriment of the others. Hence it assumes that population growth will not reduce participation (through competition for jobs) or productivity (through competition for resources and markets).

4 *Growth rate costs nothing.* The infrastructure, equipment and professional personnel that added people need will be created without penalty. The Intergenerational Reports contain no reference to these costs, although they far outweigh the extent to which population growth can moderate ageing-related costs.

Although it is universally acknowledged that wealth is measured *per capita*, the 3Ps precept ignores the fact that additional people add equally to the numerator and denominator. Hence population growth can only improve wellbeing if it increases participation or productivity, or improves wealth distribution. Interdependence of the factors is thus assumed, even though the formula implies their independence. We will examine the evidence for these relationships.

Demand and supply of labour

This brings us to analysis of the second assumption, that job seekers create jobs. The reasoning given is that people of different ages have different levels of participation in the workforce and, by increasing the proportion in the key working age cohorts, we can increase the amount of work done and thereby the wealth of the nation and the revenue of government.

Both Treasury, in the intergenerational reports, and the Productivity Commission, in its 2005 report 'Economic Implications for an Ageing Australia' assume that the proportion of people working in any age group will be unaffected by the number of people looking for work (Productivity Commission 2005). The Productivity Commission justifies this assumption by stating 'Unemployed people and people outside the labour force are generally different from the employed in skill, motivation and aptitude.' In its recently published report 'An Ageing Australia', the Productivity Commission (2013) reaffirms its belief that 'population ageing reduces aggregate participation rates' despite noting the current trends of increasing participation of older people in the workforce, and increasing educational attainment (associated with more sustained workforce participation).

This conclusion frames the context of the government's strategy. It states that the workforce is constrained by the supply of workers, thereby assuming that both capital and consumer demand are abundant to provide work for all who offer themselves. It is the basis of the 'blame the victim' approach to unemployment, welfare-to-work programs and job-readiness training.

Yet this fundamental assumption remains untested. In relation to the Productivity Commission's justification, it would be surprising indeed if employers showed no selectivity in terms of skills, motivation and aptitude, leaving those of equal employability on the shelf.

The real world experiment

The easiest way to test whether labour supply is limiting economic activity would be to compare levels of employment in comparable countries with differing levels of ageing. If a falling proportion of people of 'working age' correlates with a falling proportion of the total population in employment, this would support the conclusion that the supply of workers is a limiting factor. If there is no such trend, it implies that the supply of jobs is limiting.

Luckily we have just such an experiment playing out in the real world. Japan and Germany have almost twice the old-age dependency ratios as Australia, USA and Canada – the most youthful developed nations. Other comparably wealthy countries lie between.

Figure 6.3A shows that the proportion of the total population that is employed varies little among these nations, and does not fall with demographic ageing. If we further take account of the different proportion of part-time work in each country, the number of full-time equivalent jobs per head of population is even more uniform. Spijker and MacInnes (2013) further note that the proportion of the total UK population employed has barely changed in the past 60 years, although 'old-age dependency' has increased by half. The robustness of the proportion of people employed across demographically different nations suggests that employment is more limited by demand for labour than by its supply.

The relationship between population growth rate and growth in gross national income per capita shows equally little trend (Figure 6.3B). This offers no evidence that population growth improves productivity.

Perhaps more important than average income per capita, income inequality is a crucial determinant of societal wellbeing. A widely acclaimed study by Wilkinson and Pickett (2009) demonstrated that income inequality in developed nations is strongly correlated with worse physical health, mental health, drug abuse, education, imprisonment, obesity, social mobility, trust and community life, violence, teenage pregnancies and child wellbeing. Figure 6.3C shows a clear trend: the most youthful nations have the poorest poor. A similar trend is found using the GINI index of income inequality.

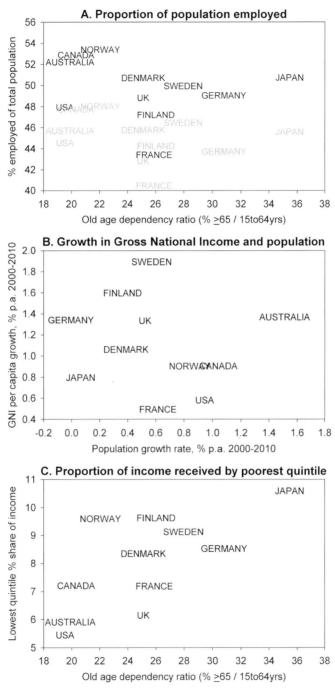

Figure 6.3: The Real World experiment: **A**. The proportion of total population employed (black), and the proportion of full-time-equivalent employment (grey), compared with the old-age dependency ratio for a range of wealthy countries. **B**. The relationship between growth in gross national income per capita and population growth rate, for the same nations. **C**. The relationship between old-age dependency and income share of the poorest quintile of the population. (Data from World Bank's online databank.)

This is to be expected with an oversupplied labour market, and provides additional evidence of the extent of labour oversupply. As the Productivity Commission (2011, p. 7) stated, 'Because immigration makes labour more abundant relative to the existing stock of capital and land, it tends to increase the returns to the latter at the expense of labour.' In addition to depressed wages, youthful societies suffer elevated unemployment and underemployment, and high cost of living due to high housing and utility costs and poor infrastructure provision.

How will we afford the pensions?

If the proportion of people without work does not increase with ageing, the burden of social transfers is unlikely to increase much either. Extra old-age pensions would replace unemployment benefits or disability pensions along with a proportion of family payments, with the benefit that fewer working-age people are excluded from the workforce.

Many nations have already responded to ageing by scheduling increases in the pension age. However, even this measure is unnecessary if labour markets are oversupplied. Unless the proportion of people employed actually begins to shrink, forcing people to delay retirement only prevents a younger person from getting a job. Even if employment were to shrink as a result of ageing, this may represent an increase in productivity, age-specific income and societal wellbeing as the least necessary and rewarding jobs are shed.

Health care is a growth industry

Most increase in health costs is due to changing treatment technologies and expectations (Productivity Commission 2013). An ageing population certainly has an increasing need for health care, but it accounts for little of the recent escalation in health spending.

The cost of health care does not increase in proportion to the number of retirees. Zweifel *et al.* (2004) demonstrated that most cost is related to proximity to death, rather than to age. Sanderson and Sherbov (2010) found that the proportion of people with less than 15 years of life expectancy is projected to increase at about half the rate of old-age dependency. The proportion of adults with disability increases even less (Figure 6.4).

However, the death rate will increase with ageing, and with it the total health care bill. Our current historically low death rate is another temporary anomaly of the demographic transition. Future death rates will still be lower than at most former times, since longevity is increased. In a stable population, the percentage of people who die each year would be 100 divided by the life expectancy at birth. We can expect the death rate to roughly double to around 1.15 per cent per year, as we

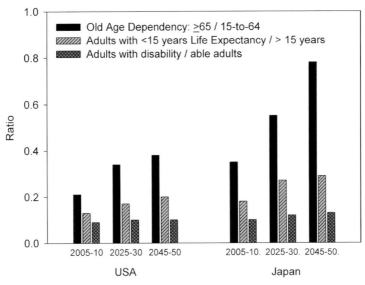

Figure 6.4: Alternative measures of ageing, comparing the change in ratio of the adult population aged 65 and over to that having a life expectancy less than 15 years, and the ratio of adults having a disability. While 'old-age dependency' will increase rapidly in coming decades, the proximity to death (relevant to health care burden) and disability (relevant to residential care burden) increase to a far smaller extent. (Data from Sanderson and Sherbov 2010.)

complete the demographic transition. We should therefore plan to roughly double the number of hospital beds per capita.

There is an obvious cost associated with increasing the capacity of hospitals and medical centres. However, the rate of hospital construction will be lower if the population is allowed to stabilise than if it is growing strongly. A stabilising but ageing Australia will need around 1480 additional hospital beds per year to match capacity to the death rate. However, if our population grows at 1.4 per cent per annum, we would need 1660 new beds per year. The extra 180 beds would cost around half a billion dollars per year.

Double standards abound in the economic treatment of ageing and population growth. Growth in health and aged care sectors is presented in public discourse as a burden to society, while construction of houses and infrastructure to cater for population growth is presented as valuable economic stimulus. Construction doesn't improve the quality of life of existing residents who are already housed. Indeed, the extra public infrastructure costs them dearly through taxes, utility charges, rates and tolls, and the extra demand for land increases the mortgage and rent burden for all.

In many ways, growth in the health industry is preferable to growth in the construction industry. It provides regular, secure and safe work that can be sustained until retirement. Much work in construction is short term and of a physical nature which forces early retirement or career shift of many tradespeople. Health services improve the quality of life for existing residents, without destroying natural amenity. Far more energy and non-renewable resources are consumed per dollar spent in the construction sector. Thus a shift in economic contribution from construction to health industries would lower the environmental impact of the economy.

Depopulation dividends

Those countries that have stabilised and become old are beginning to realise that the benefits may outweigh the negatives.

Kluge *et al.* (2014) proposed that an older and shrinking population may be:

- *smarter* – a greater proportion of people with higher education and experience
- *cleaner* – fewer greenhouse gases, as older people have lower consumption and there would be fewer people in total
- *richer* – inheritance would be concentrated to fewer recipients, rather than dissipated among many
- *healthier* – a greater proportion of life spent in wellness
- *happier* – leisure would constitute a greater proportion of the life cycle. The stresses of job insecurity and ever-increasing congestion and cost of living would also be alleviated.

When these benefits are weighed against the modest or unsubstantiated costs of ageing, it is difficult to see why nations are choosing to sustain population growth to avert ageing.

The ageing crisis we are choosing

By far the greatest threats to economic security for the aged are housing inflation and the casualisation of work. These products of excessive population growth prevent the current generation of young adults from saving for retirement. A large proportion of lifetime earnings will be spent paying a mortgage, unless they remain renters suffering ever-escalating costs of housing. The latter in particular will face a precarious retirement. The increasing proportion of casual and part-time work limits superannuation and access to credit. It parallels a shift in the burden of job training from employers to employees, with an ever-greater investment needed to gain qualifications, which are often less necessary for performing the job than for securing it. The frequency with which people change jobs has also increased, with many experiencing multiple periods of unemployment and costly location moves, eating into savings.

This is a generational time bomb imposed by current population growth.

In contrast, in stable populations like Germany, people retire with considerable savings, have modest living costs and give more to the next generation than they receive from them. Undiluted inheritance provides significant economic security for each new generation. Public investment steadily increases the standard of infrastructure, once relieved of the burden of expanding its capacity. Retirees' patronage of the arts, local tourism and recreational facilities increases the availability of these diversions for working-age people, whose own less frequent patronage would be insufficient to support the density and variety on offer. This vibrancy of a stable population is inclusive and focused on quality of experience. In contrast, the vibrancy so often associated with a rapidly growing population is characterised by crowded marketplaces with more sellers than buyers, where recreation is something reserved for elites and foreign tourists.

Just another Millennium Bug?

In the final years of the 20th century, fear spread that the turn of the millennium would throw computer date systems into confusion, bringing global financial transactions and other services to a crashing halt. Costly measures were taken to avert this danger, despite more considered advice that the fears were largely unfounded. Like the Millennium Bug, the trigger conditions for the 'ageing crisis' are inevitably reached. Like the Millennium Bug, the dire consequences simply fail to materialise. But where the measures taken to avoid the Millennium Bug were harmless, those taken to avoid ageing are daily diminishing the global prospects of achieving food security, climate stability and an end to extreme poverty.

References

Australian Treasury (2002) '2002–03 Budget Paper No. 5: Intergenerational Report 2002–03'. Commonwealth of Australia, Canberra. <http://www.budget.gov.au/2002–03/bp5/html/01_bp5prelim.html>

Australian Treasury (2010) 'Australia to 2050: Future Challenges, 2010 Intergenerational Report'. Commonwealth of Australia, Canberra. <http://www.treasury.gov.au/igr/igr2010/>

Kluge FA, Zagheni E, Loichinger E, Vogt TC (2014) The advantages of demographic change after the wave: fewer and older, but healthier, greener, and more productive? *MPIDR Working paper WP 2014-003, January 2014.* <http://www.demogr.mpg.de/papers/working/wp-2014-003.pdf>

O'Neill BC, Dalton M, Fuchs R, Jiang L, Pachaui S, Zigova K (2010) Global demographic trends and future carbon emissions. *Proceedings of the National Academy of Sciences of the United States of America* **107**, 17521–17526.

O'Sullivan JN (2012) The burden of durable asset acquisition in growing populations. *Economic Affairs* **32**(1), 31–37. <http://onlinelibrary.wiley.com/doi/10.1111/j.1468–0270.2011.02125.x/abstract;jsessionid=9079E4E881757354969065CEA605CD52.d04t0410.1111/j.1468–0270.2011.02125.x>

Productivity Commission (2005) 'Economic Implications for an Ageing Australia'. Research Report, Melbourne.

Productivity Commission (2006) 'Economic Impacts of Migration and Population Growth'. Research Report, Melbourne.

Productivity Commission (2011) *Annual Report 2010–11*. <http://www.pc.gov.au/annual-reports/annual-report-2010–11>

Productivity Commission (2013) 'An Ageing Australia: Preparing for the Future'. Research Paper, Melbourne.

Sanderson WC, Sherbov S (2010) Remeasuring ageing. *Science* **329**, 1287–1288.

Spijker J, MacInnes J (2013) Population ageing: the timebomb that isn't? *British Medical Journal* **347**, f6598.

United Nations Population Division (2011) *World Population Policies 2011*. United Nations Department of Economic and Social Affairs, New York. http://www.un.org/en/development/desa/population/publications/pdf/policy/WPP2011/wpp2011.pdf

Wilkinson R, Pickett K (2009) *The Spirit Level: Why More Equal Societies Almost Always Do Better*. Allen Lane, London.

Wilmoth J (2013) Press briefing upon publication of World Population Prospects: The 2012 Revision. Statement by Director, Population Division Department of Economic and Social Affairs, United Nations Thursday, 13 June 2013, UN Headquarters, New York. <http://esa.un.org/wpp/Documentation/pdf/WPP2012.press.briefing_Directors.remarks.pdf>

Zweifel P, Felder S, Werblow A (2004) Population ageing and health care expenditure: new evidence on the 'red herring'. *The Geneva Papers on Risk and Insurance* **29**(4), 652–666.

7

The propaganda campaign against peaking fossil fuel production

Michael Lardelli

Energy and civilisation

A formal definition of 'energy' is 'the capacity to do work'. The overwhelming majority (~80 per cent) of the work done in our advanced technological society (i.e. 'economic activity') is done using the energy released by burning fossil fuels. In fact, even a large part of the work done by humans themselves can be attributed to fossil fuels since 30 per cent of all fossil fuel use is for growing, processing, distributing and cooking the food that powers human bodies (Aleklett 2010). Of course, food production is vital when considering the future of our nation and of world civilisation.

The act of people living in cities (the origin of the word 'civilisation') is only possible when farmers produce food surplus to their own needs. Only then can humans gather in concentrated groups to do something other than farming. Two thousand years ago the world's then largest city, Rome, commanded an empire powered solely by the solar energy captured through agriculture, silviculture and fishing. It is sobering to remember that, without fossil fuels, the agricultural practices of that time produced such a small surplus that Rome's 'foodshed' spanned the entire Mediterranean region and beyond. It is difficult to envisage how today's swelling cities and national populations of many millions can be supported without fossil-fuelled agriculture. For this reason it is important to understand possible future rates of fossil fuel production.

Future rates of fossil fuel production

It is energy per unit of time ('power') that drives economic activity and permits civilisation. Therefore, when considering future energy availability, we must focus on possible rates of energy use rather than the amounts of energy available. When considering renewable forms of energy such as solar radiation it is intuitively obvious that an upper limit exists to the rate at which this can be captured. However, it is not obvious that limits exist to the rates at which fossil fuels can be exploited. Nevertheless, this is what is observed. Maximum rates of fossil fuel production ('peaks') are the natural outcome of a human economy functioning within limits imposed by geology, by the varying abundance and energy profitability (energy return on energy invested) of different fossil sources of energy and by technological factors. This can be understood within the framework of 'biophysical economics' as defined by Hall and Klitgaard (2012) but will not be discussed further here.

The most famous example of a limited fossil fuel production (exploitation) rate is the prediction by M. King Hubbert of an upper limit to petroleum (crude oil) production in the USA. In 1956 Hubbert used an estimated range of the oil that could ultimately be produced from the USA's geological sources to predict a range of possible dates and maximum rates of oil production. While derided before the event, Hubbert's modelling ultimately proved correct and the USA saw a maximum production of close to 10 million barrels per day (Mb/d) in 1970. Interestingly, in 1974 Hubbert applied his rather simplistic approach to estimating world oil production and saw a peak occurring around 1995 (Hubbert 1974). In recent years, many have sought to use this inaccurate published date to cast doubt on all attempts to forecast peak oil. However, Hubbert himself explained in an interview televised in 1976 that the constraining of oil production by Saudi Arabia during the 1970s' oil embargos would delay the peak by some years ('might extend the middle 80% (of all oil use) by about 7 or 8, 10 years maybe') (Hubbert 1976). That would put the peak of oil production closer to the year 2006 which is when the International Energy Agency admits that production of conventional crude oil plateaued.

The first scientific article on peak oil was published 10 years ago by Kjell Aleklett and Colin Campbell (Aleklett and Campbell 2003). In it they estimated that the rate of oil production in the period spanning 2009 to 2012 would be around 85 Mb/d (Figure 7.1). A decade ago the International Energy Agency (IEA)

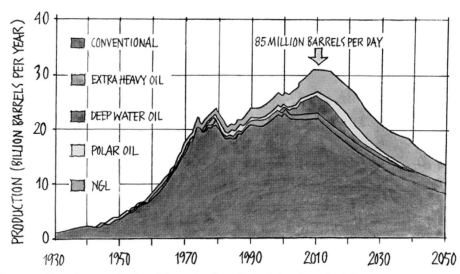

Figure 7.1: Above is a reproduction of Figure 11.4 from the book *Peeking at Peak Oil* (Aleklett 2012) that summarises 10 years of scientific research on peaking fossil fuels by Kjell Aleklett and his research group at Uppsala University. This figure shows the prediction for future oil production made in Aleklett and Campbell's 2003 paper in the journal *Minerals & Energy – Raw Materials Report*.

published scenarios anticipating that oil production would reach well over 90 Mb/d. We now know that Aleklett and Campbell were correct while the IEA was not. The 'fracking' boom currently underway in the USA will delay the decline in the rate of total world oil production for possibly up to five years but global oil production will certainly be in decline by 2020.

The March 2013 report released by Werner Zittel *et al.* (*Fossil and Nuclear Fuels – The Supply Outlook* from the Ludwig Boelkow Foundation in Vienna) is an update of a previous report on fossil fuels from 2008 that was noted for its pessimistic view of future world coal production. Since then a number of peer-reviewed scientific papers have come to similar conclusions. Zittel *et al.* currently see oil production at peak levels and expect declines within a few years. Production of natural gas is expected to peak around 2020 while coal production should peak within the next decade. Therefore, all forms of fossil fuel production should be in decline by the mid-2020s. Since the energy profitability of all fossil fuel production is steadily falling (see below) this means that peak net energy production from fossil fuels will be reached considerably earlier. Indeed, this is probably a fundamental contributing factor to the world's current, extended financial crisis (since, in the absence of increasing net energy production, economic activity cannot grow despite easy availability of money through low interest rates and 'quantitative easing'). Indeed, research by Aleklett's group showed that total energy output from coal in the world's largest coal province, the USA, has been flat since 1990 (Höök and Aleklett 2009).

Net energy production

In order to exploit fossil fuels, energy must be used to mine, process and distribute them. Energy is also required to build and maintain the physical infrastructure that allows this and to train, feed and house the humans that design and use the infrastructure. The energy produced from fossil fuel exploitation must exceed these energy inputs if there is to be an energy profit (net energy) to drive other activities of human civilisation.

In order to maximise profit from investment, humans consistently exploit those resources that have the lowest cost of production first. For example, the most accessible coal with the highest energy content is mined before less accessible also less energy-dense grades. As these resources deplete progressively less accessible/poorer grades of coal are mined but these require progressively more energy to be exploited. Consequently, as production of a fossil fuel peaks, the net energy from exploiting that resource is already declining and the subsequent rate of net energy decline is much more rapid than the rate of decline in fuel production. Since fossil fuels currently account for approximately 80 per cent of world primary energy production and since construction of much renewable energy infrastructure is subsidised by energy from fossil fuels, a rapid rate of net energy decline from fossil fuels would lead to rapid economic contraction (economic collapse).

Threats to the fossil fuel industries and their responses

The two major threats to the continued viability of the fossil fuel industries are decreased public demand for their products and decreased ability to supply fossil fuels.

The main threat to demand for fossil fuels is public concern over climate change due to carbon dioxide emissions. As revealed by Suzanne Goldenberg in an article in *The Guardian*, 'Secret funding helped build vast network of climate denial think-tanks' (14 Feb 2013), over $100 million was channelled to anti-climate science groups between 2002 and 2010 from wealthy conservatives in the USA. This included funds from oil industry actors such as ExxonMobil and Charles and David Koch. As Giles Parkinson wrote after the release of the International Energy Agency's (IEA's) World Energy Outlook (WEO) report for 2012:

> Basically, the WEO data suggests, there are a trillion reasons for the global coal lobby to resist change. That's one trillion dollars each and every year – the loss in annual revenue for the coal industry if the world takes serious action to prevent global warming, rather than just continuing on in business-as-usual.

An imminent decline in the ability to 'produce' fossil fuels (following peaks in production) threatens not only the future income of fossil fuel companies but also their ability to find investment funds for future production. According to the IEA's 2011 WEO report, almost $20 trillion must be invested in oil and gas energy supply infrastructure between 2011 and 2035 to ensure supply. That is approaching $1 trillion per year. Governments also fear the concept of peak oil since a belief in future scarcity might wipe out confidence in economic growth, disrupt stock markets and cause economic contraction before the inevitable contraction that reduced energy production itself would bring.

The main tactic used by the fossil fuel industry to avoid discussion of peaking production is to focus on the size of apparent resources. The idea the industry promotes is that remaining fossil fuel resources far exceed those that have already been consumed. While this is debatable it is also misleading since peak oil/gas/coal refers to peak *rates* of production, not the size of remaining resources or of apparently economically producible reserves.

A rather tragic example of the propaganda campaign to suppress the idea of peaking oil production is a discussion paper published in June 2012 by Leonardo Maugeri of the Belfer Center for Science and International Affairs under the moniker of Harvard University's John F. Kennedy School of Government. Maugeri's paper, 'Oil: The Next Revolution: The Unprecedented Upsurge of Oil Production Capacity and What It Means for the World', dismissed peak oil concerns and predicted a possible future glut of oil production. Unfortunately (for the author and Harvard University) the paper is riddled with simplistic and serious arithmetical errors and confusions of terms. These are well summarised in an article by the English energy journalist David Strahan, 'Oil Glut Forecaster Maugeri Admits Duff Maths' (Strahan 2012). Nevertheless, the publication online of Maugeri's paper without peer-review led to widespread and rather triumphant dismissal of peak oil concerns. Most famously, the British environmental columnist George Monbiot responded to Maugeri's paper with an opinion piece, 'We were wrong on peak oil. There's enough to fry us all: A boom in oil production has made a mockery of our predictions. Good news for capitalists – but a disaster for humanity.' Maugeri's deeply flawed report remains available at the Belfer Center website.

Inaccurate energy forecasts by the IEA

The main actor opposing peak oil concerns is the IEA, which was established in 1974 by the OECD nations to advise them on energy policy (after the oil supply crises of 1973). Despite finally admitting that the peak of conventional oil production occurred in 2006, the IEA continues to provide scenarios for increasing world oil production in coming decades based on unrealistic assumptions of rates of future oil discovery and production. Re-analysis of the IEA's data with imposition of realistic (but still very optimistic) discovery and production rates shows a future of decline rather than increase (Aleklett *et al.* 2010). The ongoing decline in conventional oil production is currently masked by a recent minor upsurge in production of unconventional oil such as shale oil in the USA. However, since conventional oil represents the great bulk of current crude oil production – and will for decades to come – it is only a question of when the decline in conventional production overtakes incremental increases in unconventional oil. An analysis by geologist David Hughes predicts that production of US shale oil should rapidly decline before 2020 (Hughes 2013).

The IEA's failure to provide realistic assessments of future world oil production has serious geopolitical consequences since many governments defer to the IEA for guidance on future energy trends. IEA assurances of continuing increases in oil production are seen by governments as justification for not engaging in the drastic measures required to ameliorate the effects of oil supply decline, for example, as recommended by Hirsch *et al.* (2005). This behaviour by the IEA appears to be the result of political pressure from the USA. A 2009 article in *The Guardian* newspaper carried these comments from an IEA whistleblower:

> The IEA in 2005 was predicting oil supplies could rise as high as 120 m barrels a day by 2030 although it was forced to reduce this gradually to 116 m and then 105 m last year,' said the IEA source, who was unwilling to be identified for fear of reprisals inside the industry. 'The 120 m figure always was nonsense but even today's number is much higher than can be justified and the IEA knows this.
>
> Many inside the organisation believe that maintaining oil supplies at even 90 m to 95 m barrels a day would be impossible but there are fears that panic could spread on the financial markets if the figures were brought down further. And the Americans fear the end of oil supremacy because it would threaten their power over access to oil resources …

Climate change and fossil fuel peaks

Imminent peaking and decline of fossil fuel production will fundamentally influence our response to climate change in two ways. It will both limit the amount of CO_2 it is possible to put into the atmosphere and limit our ability to build the non-fossil energy infrastructure required to maintain any semblance of our current civilisation.

Numerous peer-reviewed papers have shown that limits to fossil fuel production restrict the degree of global warming possible according to current models. In Chapter 17 of *Peeking at Peak Oil* (Aleklett 2010) Aleklett discusses how warnings by peak oil scientists regarding the IPCC's climate change scenarios have been ignored. Many of the IPCC's high emission scenarios described in the IPCC's Special Report on Emission Scenarios (SRES), published in 2000, assume continued fossil fuel combustion at or above current rates for the

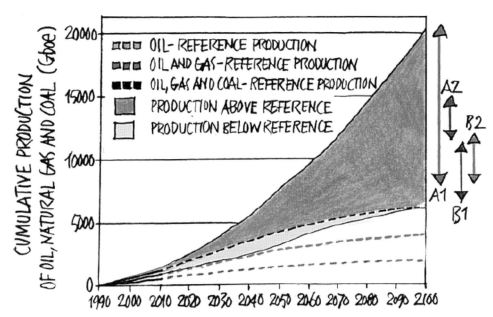

Figure 7.2: The total consumption of oil, natural gas, and coal of the 40 scenarios in the SRES until 2100 is compared with the sum of our reference predictions for oil, natural gas, and coal of 5740 Gboe. By 2100 none of the scenarios lie within our estimate and by 2050 a total of 30 are above it. The ranges of consumption variation in 2100 for the different members within the A1, A2, B1 and B2 are outside the upper limit of possible production.

rest of this century and beyond. This is simply unrealistic. In fact, according to calculations by Aleklett's group, by 2100 all of the 40 SRES scenarios would require burning of more fossil fuel than they consider it likely will be produced (see Figure 17.10 from *Peeking at Peak Oil* reproduced above in Figure 7.2 – note that 'Gboe' is gigabarrels of oil equivalent).

The relative unwillingness of the world to accept peak fossil fuel forecasts compared to the widespread (though incomplete) acceptance of the reality of climate change may be due to an arrogant and irrational belief in humans' ability to overcome any physical limits to economic growth. Thus, if humanity suffers due to an unwillingness to grapple with climate change it will have been, nevertheless, a choice we made of our own free will. In contrast, the idea that humanity's achievements might be limited by forces beyond our control – such as limits to available net energy – is currently unacceptable.

The Australian Government's failure to address peak oil

In the past decade, a number of political figures have demonstrated an awareness of the threat to Australia's food and economic security posed by peak oil. Indeed, as early as 2004 the then Australian Deputy Prime Minister, John Anderson, warned of:

... the very real prospect that at some stage in the next few short years global production may very well peak and it may be hard to increase it further at a time when countries like China, of course, are looking for a lot more fuel and even in places like Australia our dependence on oil, on petrol and transportation continues to increase.

However, in 2008 Tony Abbott, currently Australia's Prime Minister, stated he was unconcerned about Peak Oil on the basis of the neoclassical economics argument that higher prices would bring on increased supply (Abbott 2008).

The current leader of the Australian Greens, Christine Milne, was active in the Senate in 2009 pushing for an investigation into this issue. As reported in Queensland's *Courier-Mail* of 20 November 2009:

... on Wednesday Greens deputy leader Christine Milne noted in the Senate that: 'Neither the former Howard government nor the Rudd Government implemented the first recommendation of the 2007 Senate Rural and Regional Affairs and Transport Committee report into Australia's future oil supply and alternative transport fuels.'

This recommendation was that Geoscience Australia, ABARE and Treasury reassess both the official estimates of future oil supply and the 'early peak' arguments and report to the government on the probabilities and risks involved, comparing early mitigation scenarios with a business-as-usual approach.

She called on the government to 'develop a national plan to respond to the challenge of Peak Oil and Australia's dependence on imported foreign oil'. The motion was defeated 31 votes to six.

Andrew McNamara was the Minister for Sustainability, Climate Change and Innovation in the Queensland State Labor government from 2007 to 2009. He was well known for his concerns regarding peak oil and population growth although his speaking on these issues was restricted during his tenure. A conversation that I had a number of years ago with Kate Ellis, the Australian Labor Party member for the federal seat of Adelaide, indicated that Andrew McNamara had been very active in distributing information on these topics within the Labor Party. Nevertheless, the Rudd/Gillard/Rudd federal Labor government of 2007–13 took little action that could be regarded as increasing Australia's fuel and food security. On the contrary, it accelerated the already high levels of immigration that had been put in place by the preceding Howard government.

In light of the above, it came as a great surprise to those concerned with peak oil when it was revealed that a detailed draft report on peaking world oil production had been prepared by Australia's Bureau of Infrastructure, Transport and Regional Economics (BITRE) within the federal Department of Infrastructure, Transport, Regional Development and Local Government. The 436-page draft report, 'Report 117: Transport energy futures: long-term oil supply trends and projections', analysed past and forecast production for individual oil-producing nations and regions to build up a cumulative picture of world oil production (BITRE 2009). It concluded that a peak of world oil production would most likely be seen before 2020. The report was sent out for a form of peer-review by (presumably) other organisations with interests in this area in 2009. Upon receipt of the reviews, BITRE withdrew the draft report rather than modifying it for eventual release. In May 2013 I made a request under Freedom of Information legislation to see the reviewers' comments on the withdrawn report (and these are now available at the Department's website). There appears to be little in the comments that would justify abandoning the report other than from one reviewer apparently aligned with the IEA. Notably, that reviewer stated:

> *Generally speaking, the paper considers, 'The main constraints on production are geological in nature', which stands in contrast to the IEA's long-held position that it is rather above-ground factors that are likely to constrain oil supplies in the short to medium-term. … Also given its use of data, the paper would seem to be too heavily dependent on sources from 'peak oil' proponents. … Thus the paper seems to start from a given assumption, namely that 'peak oil' is inevitable, before searching for supportive evidence.*

This curious comment (that apparently rejects the finite nature of oil resources and the existence of geological constraints) should be read in the light of my earlier comments on the greater accuracy of oil production forecasts by 'peak oil proponents' and the failure of previous IEA forecasts. Indeed, one of the other two reviewers of the withdrawn report (apparently from Germany's Federal Institute for Geosciences and Natural Resources) commented:

> *I agree with you that peak production will be reached around 2020 which corresponds with our position.*

Australia's significant and increasing fuel insecurity

In February 2013, NRMA Motoring and Services released a report, 'Australia's Liquid Fuel Security', prepared by retired Air Vice-Marshall John Blackburn AO. The report warned that Australia has become increasingly dependent on imported oil since our own oil production peaked in 2000. In particular, our transport sector is overwhelmingly dependent on oil rather than other energy sources and we now rely greatly on shipments of transport fuel from refineries in Singapore (that will become increasingly dependent on oil from Middle Eastern nations). A disruption to oil flow from the Middle East (for example due to a military conflict that closes the Strait of Hormuz) could relatively rapidly have severe impacts on the delivery of fuel, food and pharmaceuticals to Australian retailers due to the low stocks of fuel held by Australia. Indeed, Australia is currently notable for holding well below its OECD-mandated '90-day net oil import stockholding obligations'.

In the longer term, Australia's agricultural production is also very dependent on oil for both diesel fuel and provision of agricultural chemicals. Disruptions or simply decreases in the availability of liquid fuels to the Australian economy threaten both food production and delivery at a time when population growth is rapidly increasing our domestic consumption of Australia's agricultural surplus. Indeed, statistics from the Australian Bureau of Agricultural and Resource Economics (ABARE) clearly show that, in drought years (such as 2006/7) Australia only produces enough grain to cover 1.6× current consumption – and that is under the currently prevailing conditions of sufficient liquid fuels (Lardelli 2010). Increasing Australia's population in the face of imminent fuel shortages and climate change cannot be a pathway to food security.

References

Abbott T (2008) Statement made at the Sydney Writers' Festival 2008 during a session titled, 'The Future of the Liberal Party'. <http://www.youtube.com/watch?v=tCiHFyLIfu8>

Aleklett K (2010) Comment by Kjell Aleklett at <http://aleklett.wordpress.com/2010/05/19/agriculture-as-provider-of-both-food-and-fuel-2/> referring to a recent publication: Johansson K, Liljequist K, Ohlander L, Aleklett K (2010) *Agriculture as Provider of Both Food and Fuel.*

Aleklett K (2012) *Peeking at Peak Oil.* Springer, Heidelberg.

Aleklett K, Campbell C (2003) The peak and decline of world oil and gas production. *Minerals and Energy – Raw Materials Report* **18**, 5–20.

Aleklett K, Höök M, Jakobsson K, Lardelli M, Snowden S, Söderbergh B (2010) The Peak of the Oil Age – analyzing the world oil production Reference Scenario in World Energy Outlook 2008. *Energy Policy* **38**, 1398–1414.

BITRE (2009) A copy of the draft report is available at: <www.manicore.com/fichiers/Australian_Govt_Oil_supply_trends.pdf>

Hall C, Klitgaard KA (2012) *Energy and the Wealth of Nations: Understanding the Biophysical Economy*, Springer, Heidelberg.

Hirsch R, Bezdek R, Wendling R (2005) 'Peaking of World Oil Production: Impacts, Mitigation and Risk Management'. Report for the US Department of Energy, February, 2005.

Höök M, Aleklett K (2009) Historical trends in American coal production and a possible future outlook. *International Journal of Coal Geology* **78**, 201–216.

Hubbert KA (1974) 'Oil, the Dwindling Treasure' M. King Hubbert *National Geographic* (June).

Hubbert KA (1976) 'Health Facilities and the Energy Crisis: A Conversation with M. King Hubbert'. Videocassette. Chicago: American Hospital Association, 1976. Available at Southwest Minnesota State University library, Media VT98. Available also at National Library of Medicine, Bethesda, Maryland, General Collection, WX 165 VC no. 6 1976.

Hughes JD (2013) 'Drill, Baby, Drill'. <http://www.postcarbon.org/drill-baby-drill/report>

Lardelli M (2010) 'Can we feed a "Big Australia"' originally published by *Energy Bulletin*, 6 May 2010. Archived at <http://www.resilience.org/stories/2010-05-06/can-we-feed-%E2%80%9Cbig-australia%E2%80%9D>

Strahan D (2012) 'Oil glut forecaster Maugeri admits duff maths'. First published in *EurOil*, 24 July 2012. Available at <http://www.davidstrahan.com/blog/?p=1570>

8

The coming radical change in mining practice

Simon Michaux

Something is happening around us. It's been highly visible for five years or so by those who choose to look. It seemed that Australia had missed the troubles plaguing the United States and Europe. The global financial crisis (GFC) seemed to not bother us here at all. Australia was doing well, largely due to the economic performance of the mining industry in a boom cycle. But now the party seems to be over.

That mining boom has clearly moved into a contraction cycle. The mining industry has seen mass lay-offs and large operation shutdowns, resulting in troubled economic predictions for the Australian economy. Demand for iron ore and coal in particular has dropped. Existing operations are struggling to avoid being shut down. There is no investment appetite for new or greenfields projects. Most gold producers have heavily over-borrowed on the assumption that gold demand will rise dramatically on the back of international geopolitical instability and currency uncertainty. While this was not a bad assumption, the gold price has dropped sharply, catching many operations overextended.

Many large corporations that invest in heavy industry and mining have recently assembled their financial analysts and optimised all their operations for

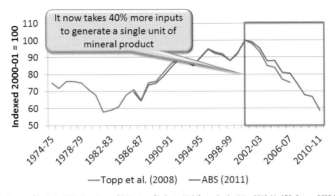

It now takes 40% more inputs to generate a single unit of mineral product

—Topp et al. (2008)   —ABS (2011)

Australian Bureau of Statistics 2011, *Experimental Estimates of Industry Multifactor Productivity, 2010-11*, ABS, Cat no: 5625.0.55.002, *Canberra*.

Figure 8.1: The Productivity Index – the efficiency in which capital, labour, materials, services, and energy are utilised to generate a unit of product.

economic performance in a risk-averse market. This has resulted in the decision to shut down many operations as they had become too marginal in a contracting global economy. Those shutdowns are still in progress.

Economies of scale have been used to mine and process more ore to meet demand targets. In spite of this, increased inefficiency can be shown across the mining industry, where the Productivity Index for mining has fallen by 40 per cent (Figure 8.1).

This means it now takes 40 per cent more work to extract the same unit of metal from the ground compared to 10 years ago. Investors are now much more aware of risk. Mining requires long-term investment, where return on capital could take 10 years. Since the GFC, investors are much less confident in the current global economy around them.

Mining operations now are at such a large scale that billions of dollars are required to start an operation, which requires an investment support that is historically unprecedented. The scale of mining operations is now at a level that has never been seen before.

Mining is not as viable as it used to be – why?

Mining is no longer the financial bonanza it used to be. There are a number of technical reasons for this, which have exacerbated other global scale issues (financial and geopolitical) and which could be seen as a marker for fundamental change as to how our industrial sector functions. These include:

- decreasing grade
- increasing rock hardness
- higher strip ratio

- increase in penalty elements (presence of arsenic or cyanide, etc. in the final product concentrate from the mine which incurs a reduction in sale price when being sold to the metal refinery smelter)
- increase in required energy versus peak energy production
- decreasing grind size
- increase in required potable water
- much greater environmental impact.

Some of these issues are discussed below.

Decreasing grade

The primary driver of the trend shown in Figure 8.2 is the lowering grade of deposits. All of the high grade, easy to work and easy to access deposits have been mined out first, leaving the difficult low grade deposits in increasingly remote and difficult locations (Giuro *et al.* 2010). Figure 8.3 shows the decrease in ore grade of several minerals over the last 150 years or so. Figure 8.4 shows the increase in production demand for minerals over a similar time frame.

Grade has decreased but production has increased for all minerals and metals. Because of the trend of decreasing grade, economies of scale need mining operations to double and triple in size for the next generation to be economically viable. All future operations looked at now are huge low-grade deposits, with penalty minerals more prominently present in deposits that prevent efficient

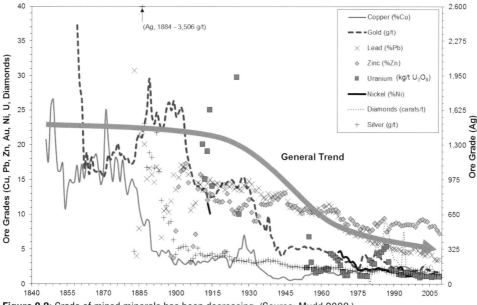

Figure 8.2: Grade of mined minerals has been decreasing. (Source: Mudd 2009.)

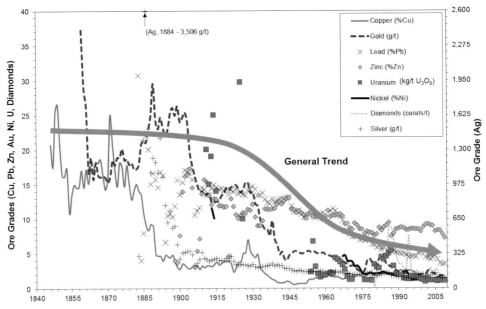

Figure 8.3: Production trends in various minerals. (Source: Mudd 2009.)

processing at ever-decreasing grind sizes. The scale of these low-grade operations will be much larger than what is done now. One example is an open super pit where four million tonnes a day are blasted and transported. Open pits are now planned to be up to 2 km deep. Some planned underground block cave operations are the size of existing open-pit operations, all of which are planned on the assumption that there will be unlimited and cheaply available energy and potable water. Also, there is the problem of the scope of stated mining reserves; these reserves are not as robust as many professionals in the industry prefer to believe (Frimmel and Muller 2011).

Increase in required energy versus peak energy production

All of these future planned operations are based on the assumption that there is an abundance of low-cost energy and available potable water. Energy consumption in mining has increased by 450 per cent in the last 40 years. This is a problem as the majority of our energy is sourced from non-renewable natural resources like gas, coal and oil.

Peak gas

The largest demand for energy is in the form of gas. Most mine sites have an electrical power station that is supplied with a gas pipeline. Conventional global gas supply would have peaked approximately in the year 2011. Unconventional gas supplies like coal seam gas (CSG) and shale gas have pushed this peak of gas production back to approximately the year 2018 (Zittel *et al.* 2013). It is now thought that, without CSG and potential shale gas supply, existing and near future projected demand for gas would not be met.

Total global gas production, including unconventional sources like CSG and shale gas, will peak approximately in the year 2018. At this time, a peak in gas production would significantly impact the industrial sector on a global scale. Without CSG, existing contracts would soon not be able to be met. Perhaps this is why CSG is being supported when it is detrimental to do so. What this means though is that the resources of arable land for food production and artesian well water have been compromised for an energy source, at a time when there are probable food and drinking water shortages. The question of *Cui Bono?* needs to be addressed to understand how this happened (Ruppert 2004).

Peak oil

The next largest consumption requirement in the mining industry after gas is diesel fuel (derived from oil). Some of the larger mines consume as much as one-third of the energy they require in transporting ore or waste from the excavating face out of the pit.

Oil is also a finite natural resource which is not being renewed as it is being consumed. The majority of oil was discovered in the 1950s and 1960s. Since then discovery has declined (Morse and Jaffe 2001). Production on the other hand has steadily increased. Currently, for every barrel of oil we discover, we are consuming between three and five barrels. It is now accepted by the International Energy Agency (IEA) that conventional crude oil production peaked in the year 2006 (IEA 2006; Zittel *et al.* 2013).

After steadily rising for decades, oil production remained static between approximately 2005 and 2008, but the selling price almost tripled in that same time. During this time, the demand for oil and supply separated, causing a shortage and an inelastic supply market. If we compare world spot price of oil ($USD/barrel) and oil production (Mb/d: million barrels oil per day) from 1998 and 2011, we can see a sharp change in gradient at the year 2005 that shows two distinct market behaviours. Prior to 2005, oil production was elastic in form, where demand could be met with available supply. Between 2005 and 2008, oil production was inelastic where supply and demand separated. In 2008, the GFC caused a great deal of economic hardship, which in turn reduced global demand for oil. Since the GFC, supply has been able to meet demand and that inelastic pattern is no longer in effect (but market conditions are appropriate for a return of that pattern).

It has been theorised by many analysts that, as we are a petroleum-supported economy in both a global and Australian context, this inelastic supply market for oil applied extraordinary pressure on the fiat finance monetary and fractional reserve banking systems, triggering the GFC. The weakest link in these finance systems was the US housing market, which collapsed. As we are a global system, the consequences of that collapse were transferred to every economy around the world in 2008. This set of fundamental circumstances is now occurring again in the 2013–2014 time period.

Unconventional supply sources like tar and oil sands have pushed the peak of total oil supply back six to seven years (Hughes 2013). Total peak oil was projected to be approximately in the year 2012. For the last five years, many economies around the world have been subject to economic contraction. This, among other things, has translated into the destruction of demand for oil. As a direct consequence of this, we have yet to feel the pinch of demand for and supply of oil separating again.

We are heavily dependent on oil to function as an economy. There is no clear replacement for oil when it is taken off line as it has more energy per unit volume in calorific content than any of the alternative fuel sources at this time (Hirsch *et al.* 2005). It has been observed that there is a strong correlation between the places where the proposed need for military engagements on behalf of the War on Terror have been conducted, and rich natural resources (in particular oil) where those engagements happen (Ruppert 2004). As oil reserves contract and there is no longer enough to service all demands, an expected response would be resource wars, marketed as something else.

Peak total energy

In the past, it has been difficult to compare the quantity and quality of various energy sources due to lack of hard data. Recently this has been addressed (Zittel *et al.* 2013). Figure 8.4 shows oil, gas, coal and uranium data normalised and plotted on the same graph. To do this, the following conversion factors were used:

- 1 Mtoe* = 7.1 million barrels of crude oil and condensate
- 1 Mtoe = 10 million barrels of natural gas
- 1 Mtoe = 1.16 billion cubic metres of natural gas liquids
- 1 Mtoe = 1.5 million tonnes of hard coal (1.8 million tonnes sub-bituminous coal)
- 1 Mtoe = 3 Million tonnes lignite
- 1 Mtoe = 58 tonnes uranium.

Total world fossil fuel supply is close to peak, driven by the peak of oil production. Declining oil production in the coming years will create a rising gap that other

---

* Million tonnes of oil equivalent

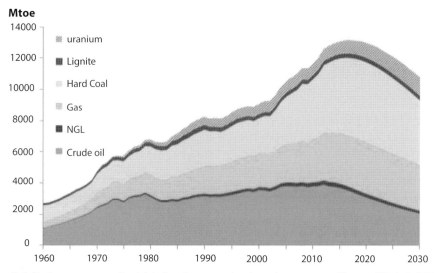

Figure 8.4: Peak energy, normalised data for oil, gas, coal and uranium reserves. (Source: Zittel *et al.* 2013.)

fossil fuels, like gas or coal, will be unable to compensate for. The energy contribution of nuclear fuels is too low in order to have any significant influence at a global level, though this might be different for some countries. Moreover, as with fossil fuels, easy and cheap-to-develop mines are also being depleted. The effort and cost of uranium production will continuously increase as a consequence.

The industrial systems that each of these energy sources supports are quite different and are not interchangeable easily. That being said, each of those industrial systems is vital for our society to function. Putting all energy sources together gives a snapshot of our industrial capability. Peak total energy is projected to be approximately in the year 2017, three years away. As all of these sources are only a few years away from peaking and declining (with the exception of uranium), a compelling case can be made that our society and its industrial sector energy supply face a fundamental problem, a problem that is systemic in nature.

The Big Picture

Clearly these trends cannot continue. This implies that a radical change in practice is coming to the mining industry, the manufacturing sector it supplies and to the global society that it supports. The issues discussed in this article are not just the coming change in practice for mining but also our society's ability to function at all in an industrial context due to peak energy.

Put these observable trends together and a compelling case can be made that our society is approaching an existential crisis that is systemic in nature and that it is in denial about the existence of that crisis (Figure 8.5).

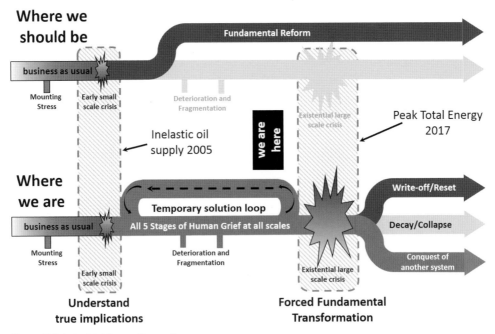

Figure 8.5: Industrial flow path over time.

The basic pattern set out in Figure 8.5 can model how we react globally to systemic problems and what we should do about them. For our industrial grid, energy supply is the decisive risk for continued operation in its current form. The estimated time for peak energy is 2017. This is the existential crisis for our industrial sector. All industrial manufacture will be forced to transform into something else once a permanent decline in available energy manifests. It is the author's opinion that the early crisis was the inelastic oil supply pattern seen in 2005, which was resolved by a global economic crash (the GFC). We are currently trapped in a repeating loop, as we keep on doing the same things. We are doing this through the printing of money to meet finance targets and through shutting down operations. This is the plan until economic demand can sustain the previous growth targets. Unfortunately peak energy results in no further economic growth in macro terms.

Eventually we will be forced though the existential crisis point of an energy crash only after a series of economic crashes. As there is no clear Plan B, we will not be able to simply write off and reset. It is the author's opinion that when we get to this point, the industrial sector will decay and collapse in a non-linear non-homogenous fashion. Regions and various industries will continue at the expense of other industries and regions. Some parts of the world will try their level best to maintain the existing paradigm, even if it means cannibalising their own marketplace, in the hope that it will all get better soon. Other parts of the world will embrace fundamental change while the existing system still works and survive.

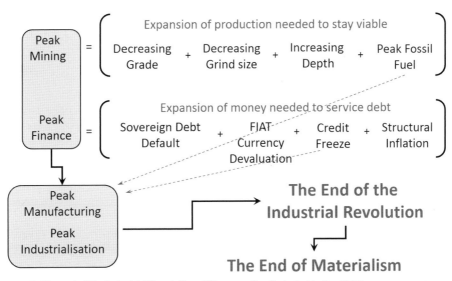

Figure 8.6: The end of the Industrial Revolution, 160 years after it started in the 1850s.

It is very clear, though, that our industrial society is about to go through a permanent change. Much of what is around us now is utterly dependent on manufactured goods and the supply system to deliver them. Mining is the industry that sources the raw materials. While the time lag for anything like a copper/iron/aluminium/metal shortage is far longer than for say a financial crash or an energy shortage, the implications of a supply/demand gap in mining have much more far reaching consequences. Figure 8.6 shows *some* of the knock-on consequences of this combination of issues.

Current trends in mining require it to expend more and more energy at higher capital cost for the same unit of metal as time goes on. There is a need to expand in scale to remain viable. Peak energy, however, means that this is not possible and so the rate of mining will peak and decline. These industrial projects require industrial procurement to function safely and for it all to work effectively. There are currently a number of structural risks in the finance sector that make this difficult. The European debt crisis, the US government debt default, $USD and € devaluation and hyperinflation of all fiat currencies could all result in a credit freeze for industrial procurement. All of these things are possible and now even probable at some point.

Peak raw materials (peak mining) and peak finance result in less manufacture of goods (peak manufacture). Peak manufacture combined with peak energy and a credit freeze in industrial procurement result in peak industrialisation. There will be no more Hoover Dam or Snowy Mountain Hydro projects, because we simply won't be able to logistically do them. All of this combined means the end of the Industrial Revolution, 160 years after it started in the 1850s. From a social point of view, this would mean the end of materialism as, not only would the widespread manufacture of goods contract sharply, money to purchase the goods would also be unavailable.

The writing on the wall for us all

Everything we need and want to operate in our current way of life is drawn from non-renewable natural resources in a finite system. Most of those natural resources that we need are depleting or will do so soon. Conversely, demand for everything we need and want is expanding fast in the name of economic growth (and increasing population). When these trends meet, there will come a point where the way in which we do things will fundamentally change.

None of these issues can be seen in isolation. Everything interacts. This means that a chain reaction is probably what is going to happen. A traditionally isolated problem will happen, which will trigger unprecedented chaos. This is the nature of systemic crises around fundamental support services.

Traditionally issues like peak energy or climate change have been discussed in isolation, where each issue could be seen as an enormous challenge to our society. What is actually happening is that multiple macro-scale problems are reaching a civilisation-wide pain threshold all within a five to 10-year time period.

It is the author's opinion that we will see a systemic crash of the fiat currency financial system first. Fiat currencies spiralling into hyperinflation and the resulting consequences are well understood as there are multiple historical examples to reference our current circumstances against. Historically, a currency default has followed a financial debt reset, and the new monetary system is able to grow after some upheaval (often with a war or military engagement as an economic stimulant).

In the present case, this will not be possible because soon after our society will experience peak energy. This is relevant as energy translates into the ability to do work, which means that there is not enough energy to expand the existing system. So while there is a systemic financial crash in progress, there will be the need to develop and build an energy system with alternatives to oil, gas and coal (nuclear is also unviable but beyond the scope of this article). Preferably, this alternative system would supply the same quantity of energy that the current fossil fuel systems do (though highly unlikely). In addition to this, the power grid network that provides the industrial infrastructure also needs to be rebuilt to cater for the new energy system.

This is where peak mining is relevant. There will not be enough raw materials, metals or minerals to meet the demand to restructure the energy grid, even if there were the industrial procurement systems (finance) in place to pay for it.

While all of this is happening, systemic environmental breakdown will continue to disrupt any industrial development. Continued incidence of major

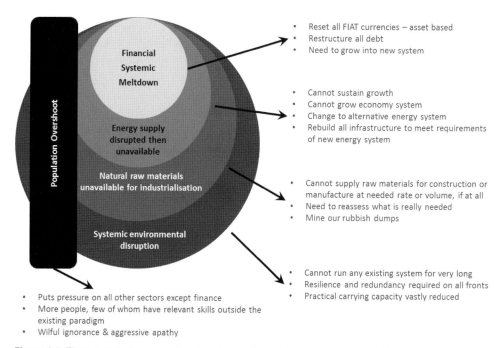

Figure 8.7: The network of dynamically interacting and exacerbating large-scale problems.

flooding, cyclones/hurricanes/typhoons/tornadoes, bush fires and droughts will make any industrial reform very difficult. This is something we have been experiencing for the last few years at greater magnitudes.

Overrunning all of these issues is overpopulation. There are simply too many people, all of whom are putting pressure on all other sectors except finance (which currently is based in confidence, not reality) (Figure 8.7).

The human population has increased and the per capita of energy consumption has increased (Foran and Poldy 2002). These two patterns together show that consumption of non-renewable natural energy resources has increased 45 times in the last 150 years, while the planet we live on is a dynamically adjusting and ultimately very stable finite system.

The rate determining problem is we are out of time to innovate any kind of solution to this network of issues.

The party is over when demand for something vital outstrips supply or when some vital service ceases to function reliably (or at all). The flashpoint is not the peak production of any given natural resource but the perception by the average person. This happens when the people en-masse understand that the world they live in is no longer possible. Once the voting public understands that there is no easy solution at hand that will allow their lives to continue in the fashion they have become accustomed to, then there will be no avoiding the supporting problems.

Failure to address these problems will result in our society being devastated. There is a school of thought that tells us that there is considerable effort to keep the voting public largely ignorant of these issues, to keep them at their posts, working and consuming. What happens to the idea of democracy when there is no longer enough to go around? If we wish to stay a democracy then the average person must become educated in these issues and actively take part in developing the solutions.

The challenge for our political leadership is considerable. A series of solutions are required and then an unprecedented amount of leadership and vision needs to be applied. The voting public has to understand what the genuine issues are and all stakeholders then have to work together. Our current approach seems to be one of wilful ignorance and 'give war a chance'.

We either meet these problems effectively, or those problems meet us with devastating consequences.

References

Foran B, Poldy F (2002) *Future Dilemmas: Options to 2050 for Australia's population, technology, resources and environment.* CSIRO Sustainable Ecosystems, Canberra.

Frimmel H, Muller J (2011) Estimates of mineral resources availability – How reliable are they? *Akademie für Geowissenschaften und Geotechnologien Veröffentl* **28**, 39–62.

Giuro D, Prior T, Mudd G, Mason L, Behrisch J (2010) 'Peak Minerals in Australia: A Review of Changing Impacts and Benefits'. Cluster Research Report 1.2, Institute of Sustainable Futures, Sydney University and Department of Civil Engineering Monash University, CSIRO.

Hirsch R, Bezdek R, Wending R (2005) *Peaking of World Oil Production: Impacts, Mitigation and Risk Management.* US Department of Agriculture.

Hughes D (2013) *Drill, Baby, Drill – Can Unconventional Fuels Usher In A New Era Of Energy Abundance?* Post Carbon Institute, California.

IEA (2006) Peak oil happened in 2006. International Energy Agency, Paris. <http://makewealthhistory.org/2010/11/11/iea-peak-oil-happened-in-2006/>

Morse E, Jaffe A (2001) 'Strategic Energy Policy – Challenges for the 21st Century'. A report of an independent task force cosponsored by the James A Baker III Institute and The Council of Foreign Relations.

Mudd G (2009) *The Sustainability of Mining in Australia – Key Production Trends and Their Environmental Implications for the Future.* Department of Civil Engineering, Monash University and the Mineral Policy Institute, Melbourne.

Ruppert M (2004) *Crossing the Rubicon – The Decline of the American Empire at the End of the Age of Oil.* New Society Publications, Canada.

Zittel W, Zerhusen J, Zerta M, Arnold N (2013) *Fossil and Nuclear Fuels – The Supply Outlook.* Energy Watch Group, Berlin.

9

Coal: nails in the global coffin

Sharyn Munro

My book *Rich Land, Wasteland* is subtitled *How Coal Is Killing Australia* because that's the reality I discovered, but the underlying truth, much-ignored in Australia, is that coal is killing our entire planet (Munro 2012).

I began locally; perhaps I had been naive in thinking that it was government's role to care for its people, but I felt driven to act, for my grandchildren's sake, when it finally sank in that the New South Wales (NSW) government was ignoring the accumulating adverse impacts from unbridled coal mining on the air and water, health and futures, of the people and places of the Upper Hunter Valley – and was planning to increase them.

In this once-rural part of the valley, where they had coped with only a few mines 'to keep the lights on' for us all, the industry boasts that the 50 km between my grandchildren's hometown of Singleton and north to Muswellbrook is taken up with open-cut mines, giant holes and bare overburden mountains, so many and so close that the area is visible from space, not a scar, but a running open sore, ever-expanding, ever-polluting.

State strategic land use plans saw the industry's expansion in this oversaturated region as unavoidable, because 'there is an ongoing demand for our coal' – 84 per cent of Hunter coal is exported. This is thermal coal to be burnt elsewhere for energy, which I would suggest is likely to be banned and phased out worldwide in the not-too-distant future, since 'clean coal' remains, as one protest sign says, 'a dirty joke' – still not commercially viable.

As the latest Intergovernmental Panel on Climate Change (IPCC) report re-confirms, man-made global warming is no joke, and it is largely fuelled by CO_2 from burning fossil fuels. We are being told by the World Bank that we must leave most of the world's coal in the ground if we are to have any chance of curbing global warming. But who in our governments will be allowed to listen to such heresy, let alone act on it?

While it is still legal, let's dig it all up, no matter what harm it causes locally and globally, and even though it didn't keep Queensland or NSW in the black – because we are locked into supplying a demand.

A demand is just a market; we actually are not obliged to keep trashing our country to satisfy a self-destructive demand elsewhere:

> Even if the spin about this industry's economic value were true, Australia should not be staying afloat economically on the pain of its people and the destruction of its life support systems for water and food. There are smarter ways to generate revenue than selling off the country, breaking it up and sending it away in fleets of coal ships and LNG tankers, for other countries to perpetuate their fossil fuel dependency and add to the demise of our world and its inhabitants. (Munro 2012, p. 424)

The Climate Institute estimated that, if Australia builds up its coal exports as so eagerly planned, we would produce 30 per cent of the carbon needed to push global warming beyond two degrees.

With nine mega-mines planned for Queensland's Galilee Basin, exporting coal through the Great Barrier Reef, the coal from there alone would create annual emissions equal to 130 per cent of Australia's current total. Yet these massive off-site emissions are not counted in environmental impact assessments. It will be enough of an environmental atrocity if Clive Palmer is allowed to mine the Bimblebox Nature Refuge there and wipe out its richly diverse flora and fauna, but what bigger impact on the environment could there be than climate chaos?

As Ian Dunlop keeps saying, we should be on an emergency war footing about global warming, but instead our Federal Government is just trying to block our ears to the sound of the approaching gunfire – for which we supplied the ammunition. Or is that the sound of cracking polar ice that I hear?

I could see that if coal was as out of control elsewhere as in the Hunter, there would be nothing left for all our grandchildren to inherit but ill-health, in a dewatered and contaminated wasteland, on a planet in climatic and probably social chaos.

What could I do about it? I only had one weapon, a way with words.

So even in my first, much more lighthearted book, *The Woman on the Mountain* (Munro 2007), I wove in concerns like sustainability and global warming and coal's role in that – and the lack of action by our 'leaders'. Sadly nothing much has changed …

> *For the overwhelming majority of politicians the issue is driven by the polls rather than a sense of urgency, with no one in power standing up to say, 'Hey, Rome is burning! Let's quit fiddling NOW!'*
>
> *They're still debating what would be the most popular tune to play next.*
>
> *They're still putting profit before the planet and its people, tinkering at the edges of the problem while trying not to inconvenience Big Business.*
>
> *After all, they don't want to be accused of being greenies. 'Sorry kids', ancient ex-pollies might say to future generations scrabbling for food in a dried-up or drowning world, 'we were too busy with Economic Progress to stop climate chaos'. (Munro 2007, p. 25)*

My sort of rich land can only exist where triple bottom line full cost accounting is respected: people, planet and profit. Our sham assessment processes for resource projects certainly don't do that.

I have watched as impacted communities and individuals, fighting to protect the health and futures of people and waterways, eventually learn that no matter how glaring the predictable damage, it will be found that the project will be 'unlikely to have significant impacts' or if any are admitted, they will be 'monitored, managed or mitigated'.

All along the coal chains, from the mines to the uncovered stockpiles and rail wagons to the port coal loaders, and now the coal seam gas (CSG) chains, from the gas fields, along the hundreds of kilometres of pipelines to the LNG plants and ports – the impact on people's lives is *shameful*.

Now the Federal Government aims to remove their oversight role, the 'green tape', the hard-won protective measures, and let the fossil-fuel-addicted, vested-interest states have their way.

Governments are not exercising their duty of care to their people or their country, and do not even *acknowledge* a duty to the planet in this coal matter.

Duty of care? The Hunter has had way over 100 air pollution exceedence alerts this year; you get email or SMS alerts, and if you have heart or respiratory conditions, you stay indoors. Singleton, where my grandchildren live, is, like Lithgow, now one of the most polluted areas in Australia, from coal and coal power. The state average for reduced lung function in children is 1 in 9; in Singleton it is 1 in 3.

Regional health departments simply record the above state average incidences of and early deaths from respiratory disease, heart attacks, strokes and certain cancers. Doctors are very concerned by the higher rate of infant mortality and foetal abnormalities.

Is this Australia or China? Australia, because, for example, China is investigating their high birth defect rates in coal areas. They can't afford not to – they don't have enough kids to waste …

Yet Singleton's poor air quality is not enough reason to reject more open cut mines there; the O'Farrell government won't even accept legal rulings, as at Bulga, or their own Health Department's advice, as at Camberwell. As if heavying the Environmental Defender's Office and removing legal aid wasn't enough, their proposed planning changes would mean that no project could be rejected or appealed against because of health or environmental impacts. They want the 'significance of the resource' to come first. Goliath gets more muscle, while they don't want David to even have a slingshot.

But even putting profit first, the balance sheet doesn't add up. Apart from wider impacts of the boom like the skewing of the economy and the high dollar, coal directly costs us money, be it via physical and mental health treatment, lost jobs in other lost industries and support services, subsidies like the diesel rebate, or the free infrastructure, like the $3.3 bn coal rail upgrade to Newcastle. Governments don't mention these.

It is not only shortsighted and unsustainable for government to pursue this coal dependence … it's not smart. The stupidity of basing our economy on a product that will shut us all down in the end is apparent to the unblinkered. Smart money *must* be considering smarter, long-term things to invest in. Coal power is not only deadly, it is dodo technology; renewables are the future – yet Tony Abbott wants to prop up coal power at all costs. Business is ahead of government on this, as they must deal in reality.

We know the price of coal has plummeted, new projects mothballed, mines closed and staff laid off. Goldman Sachs say that the era of investment in thermal coal is closing partly due to expected environmental regulations (Lelong *et al.* 2013); scientists warn that our decreasing water supplies will make coal power (and coal mining) untenable; our major public health bodies say we must urgently address the health issues of our minerals and energy policies (Climate and Health Alliance 2013); the World Bank is withdrawing from coal power funding (Colvin 2013); the Uniting Church is divesting from fossil fuel industries (ABC News 2013a).

Soon the stigma will lead to more public bodies and companies denouncing any association with harming the planet or its people. Business will have to respond, as it is already being driven by insurance and investor demands. As happened with the asbestos and tobacco industries.

An international treaty on reducing mercury emissions, recognised by the UN and the WHO as a global threat to human health and the environment, is being prepared; mercury is toxic to the nervous system and to the immune, cardiovascular and reproductive systems. Our main source of mercury emissions is coal power. Will we sign up with our left hand to reducing mercury emissions, as we did to reducing carbon, and likewise proceed with the right hand to encourage more coal mines and exports for coal power to increase the toxic emissions of both?

Many of even the most conservative people have realised that governments will continue to let them down in caring for our country and our futures. Over the last few years, as disillusionment turned to determination not to let this happen, we have seen the growth of unprecedented and often unlikely alliances.

Within them, many people, especially in rural areas, are more concerned for their local issues, like the loss of water and farmland; they do not all accept man's role in global warming; others care deeply about our carbon trajectory, and others are acting with both local and global issues in mind.

I see many reasons, social and commercial, to think a tipping point is approaching, and to be hopeful, even after 7 September 2013. In fact, the Abbott Government's swift actions to sideline or ignore so many issues related to a sustainable environment seems to have strengthened resolve to fight on, fight harder, and more strategically.

Australia is still nominally a democracy; its fall to tokenism is part of the invasion. Our governments have become dependent on the tips from the plunderers ...; while not declaring victory, the latter have all the rights of victors, since in effect they dictate to the puppet rulers. Look at the mining tax:

> ... coal-impacted and coal-aware people everywhere are ... demanding their real democratic power back – or else. The rapidly growing Lock the Gate Alliance of such people spells one form of 'or else'; organised people's defiance of bad legislation, civil disobedience to protect what our governments won't. Corporate money should not talk louder than the voices of the people, especially when they are pleading for rescue. (Munro 2012, pp. 424–5)

The environment is not at odds with jobs and the economy; it is integral. Jobs should not be at any cost. Besides, coal and gas projects do not create jobs for the unemployed, but suck workers from other industries.

To quote the late Bryce Courtenay, 'It's not about being a greenie; it's about not being stupid!' It is really because fossil fuel companies fear the rise in public understanding of this that they launched a multi-million dollar PR campaign for CSG (Daley 2013).

Anti-CSG protests in NSW and increasingly in other states had caused companies to shelve projects. But they are regaining courage, as our Federal Government urges the states to go for it all (ABC News 2013b), encouraging the myth that we must have more CSG or die – or at least the stoves and heaters will when the gas runs out while the prices go up (Crawford 2013).

It is like the old scaremongering that we would all be back in caves eating moss in the dark if the greenies made us stop burning coal. My place has been stand-alone solar powered for 20 years, yet I manage to live a most civilised life, well-catered and well-lit.

All this investment in onshore unconventional gas – coal seam, shale or tight – is locking us in for decades and delaying the urgent move to renewable energy, doable now, as the Beyond Zero Emissions plan showed (Beyond Zero Emissions 2010).

There need be no domestic urgency for us to be taking these risks, but for the facts that most of the new gas projects are committed to export, and our prices will rise to suit, and NSW imposes no domestic reserve provisions, unlike Queensland.

The greenwash sales pitch was that using methane for energy produces less greenhouse emissions than CO_2 from coal. But we know that when you include the whole lifecycle, with the 'fugitive emissions' and the fact that methane warms the atmosphere many times faster than CO_2 – although it doesn't last as long, we don't have long – its immediate impact is worse in these decades when we most need to slow global warming. The Fourth IPCC Report has predicted that by 2030 coal could account for 8.4 billion tonnes per annum (btpa) of greenhouse gases – and methane 11 btpa.[1]

My bumper sticker says, 'Coal costs the earth'. Well, so does methane.

I have been too narrowly called an environmental activist; I am really a 'common sense activist'. Common sense … like not fouling our own nest, the one planet we can call home; not aiming to keep fuelling global warming and create even more frequent extreme weather events and damage. Or not killing the goose that lays the golden eggs, à la the Great Barrier Reef and tourism; or our agricultural areas, current or potential; or any source of clean water, but especially the Great Artesian Basin, on this driest of continents, from the dewatering, depressurisation and contamination caused by both coal and CSG extraction.

China and India also have huge coal resources and suffer from a lack of water, but they have smarter governments, so it is easier to buy an Australian coal company and its mines and use our water. Indian companies are big players in the Galilee Basin, and Yancoal, majority owned by China's Shandong provincial government, is the largest single coal operator in Australia.

All Western Australia's coalmines in the southwest, on which their power stations depend, are now owned by Indian and Chinese companies. There, and in South Australia and Victoria, it is planned to treat and export poorer quality coals, like the 30 per cent more carbon-intensive brown coal, for energy – yet all the mooted treatment would only make it as 'clean' as black coal.

'Green extremists' are accused of denying the world's poor the right to electricity, as if coal power is the only way to produce electricity. Yet India's new coal power stations are in the industrial regions, not those where people have little or no electricity.

My national journey of discovery here revealed an industrial invasion of appalling scale and speed, into hitherto unthinkable places. Shortsighted, stupid, unjust, even 'insane' were the words that kept occurring to me:

> I ... wondered at the lack of connection ... as I listened to the big fuss about the Murray–Darling Basin Plan in late 2010, when farmers were being told to reduce their irrigation allocations for the common good of sustainable water use for farming and the environment, yet coal and CSG projects that place water sources at huge risk are being approved at rapid rates. I heard grand and expensive schemes for piping water across Tasmania to irrigate and create new 'food bowls'. But we already have areas with good soil and reliable water like Gloucester or ... black soil plains (like the Darling Downs) that don't need irrigating. These current and future food bowls are being destroyed by mining. How can this be anything but stupid? (Munro 2012, p. 220)

If we are to feed ourselves, let alone the world, as the population increases and global warming plays havoc with our agricultural regions, it makes no sense to be guaranteeing the worst scenario via our coal contributions.

Water and food security will become far more valuable resources than any fossil fuel; we need to be focused on increasing agriculture, not ruining arable land for coal or gas or rendering grazing land unworkable with gas fields or water loss, and on preserving all our water sources, not reducing them. Especially for an industry that runs on diesel and is water-hungry; hardly a long-term prospect:

> We ... have a right to expect that our governments will think beyond what Big Business wants and how much they will pay for it, and instead think about what will be needed for a safe and healthy population into the future, and how to achieve that. It will never be what Big Business 'suggests', no matter how green-tinged the spin that surrounds it. (Munro 2012, p. 424)

We can be so much more than the world's quarry, our futures calculated by corporate coal – and every new thermal coalmine will add nails to the global coffin. I consider it immoral to say that if we don't supply the coal, others will; drug dealers could say the same:

> It used to be said that Australia rode on the sheep's back. Do we really want to be now riding on the back of global warming, from coal ...? (Munro 2007, p. 164)

Endnote

1 See: <https://en.wikipedia.org/wiki/Natural_gas#CO2_emissions>

References

ABC News (2013a) Mining and CSG on church black list. <http://www.abc.net.au/news/2013–04-18/mining-and-csg-on-church-black-list/4636466> accessed 11 November 2013.

ABC News (2013b) Federal Government steps in to speed up coal seam gas drilling in New South Wales. <http://www.abc.net.au/news/2013–09-26/federal-intervention-to-kickstart-nsw-csg-sector/4982316> accessed 11 November 2013.

Beyond Zero Emissions (2010) Stationary Energy Plan. <http://bze.org.au/zero-carbon-australia-2020>

Climate and Health Alliance (2013) Media Release. <http://caha.org.au/media/media-releases/> accessed 10 November 2013.

Colvin M (2013) World Bank will no longer fund coal-fired power stations: PM (Transcript). <http://www.abc.net.au/pm/content/2013/s3805219.htm> accessed 11 November 2013.

Crawford B (2013) NSW to frack on with its gas plan. *Courier-Mail*, 26 January, p. 11.

Daley G (2013) Gas industry fires up with major ad blitz. *The Australian Financial Review*, 29 July, p. 7.

Lelong C, Currie J, Dart S, Koenig P (2013) The window for thermal coal investments is closing. Goldman Sachs, Commodities Research, 24 July.

Munro S (2007) *The Woman on the Mountain*. Exisle Publishing, Wollombi.

Munro S (2012) *Rich Land, Wasteland: How Coal Is Killing Australia*. Pan MacMillan, Sydney.

10

Save the soil to save the planet

Michael Jeffery

Introduction

The challenges we face in dealing with land degradation, a changing climate, food and water security, energy demands and the needs of increasing populations are unprecedented.

I believe that these challenges are interrelated and that landscape management practices, including but not limited to agriculture, forestry and fire have caused significant damage. Our precious resources of soil, water and vegetation, necessary to sustain life, are being lost globally at unsustainable rates. We have altered natural bio-systems and subsequently diminished our ability to effectively deal with these challenges.

Current situation

Despite the excellent work of innovative farmers across many regions of Australia and some good science, the reality is that this degradation is continuing. Impacts are already being felt including depletion of soil carbon and declining soil health, increasing salinity and erosion, diminishing water availability, deforestation and vegetation loss and the impacts of extreme weather events.

There are, however, innovative solutions currently being developed that can potentially reverse land degradation and equip communities to better deal with impending challenges. The answer lies in how we manage our landscape. Active management of soil, water and vegetation are the key process drivers for landscape regeneration.

Soils for Life focus on soil, water and vegetation

The objective of the Soils for Life organisation, which I chair, is to enhance Australia's natural environment through the provision of information and education on innovative leading performance in managing the landscape. Our goal is the widespread adoption of regenerative landscape management practices that will help to restore and maintain any Australian landscape to be fit for purpose.

In 2012 we launched our report, 'Innovations for Regenerative Landscape Management'. This report showcased 19 case studies of farming enterprises – across a range of regions and land-use types – which were demonstrating strong environmental and production outcomes (SoilsForLife 2012). The common theme for each of our case studies was the recognition of the need to integrate the management of their soil, water and vegetation resources.

Healthy soils

The carbon content of soil is one of the key indicators of its health and is a master variable that controls numerous processes. It is the carbon content of soils that largely governs their capacity to absorb, retain and supply moisture within the soil and to sustain active plant growth. Soil carbon helps support a healthy balance of nutrients, minerals and soil microbial and fungal ecologies, improving soil fertility.

Healthy soils promote vigorous plant growth and plant and animal resistance to disease and insect infestation. Increased soil carbon levels, therefore, also have the means to reduce our reliance on costly fossil fuels and other farming inputs.

We know that of the 39 regions assessed across Australia, only four have increasing soil carbon stocks (SOE 2011, pp. 281–286). Current rates of soil erosion by wind and water across much of Australia now exceed soil formation. Research from the CSIRO released in July 2013 indicates that Australian soils are losing about 1.6 million tonnes of carbon per year from wind erosion and dust storms alone (Chappell *et al.* 2013).

Additionally, soil acidification affects about half of Australia's agriculturally productive soils and this is, in part, attributable to current agricultural practices of excessive use of non-organic fertilisers. Acidification restricts options for land management, growing acid-sensitive crops and looms as a major constraint in Australia's capacity to increase carbon in agricultural soils (SOE 2011, p. 287).

Land management practices can either accelerate or moderate such degradation. Focusing on more than just the conventional N, P, K fertilisation programs and ensuring we are addressing the full structural, biological and mineral health of our soils has the potential to reverse the current downward trend.

Water

Soil health and water use efficiency are inextricably linked. A primary outcome of a good level of soil organic carbon is its ability to absorb moisture. Every gram of soil carbon can hold up to eight grams of water. A well-structured soil, high in organic matter and soil carbon, thus acts as a sponge, improving infiltration and retention of rainfall. This retained moisture is then released slowly for plants and animals to maintain production over a much longer period.

The benefits of this lie not only in the general sense of maintaining production, but also in maximising absorption in times of drought when rainfall levels are minimal, and also in flood, when significant rainfall should be drawn back into groundwater storage, not running overland creating inland seas.

In the future, securing an adequate supply of safe, reliable water will become a strategic determinant for communities, regions and nations worldwide – and it is imperative for our farmers to sustain production. We must be efficient in using the rainfall we get.

Currently we collect and store water in dams and reservoirs. On average, for every 100 drops of rain that fall on our landscape a staggering 50 per cent is lost to evaporation. This is 25 times the quantity stored in all our dams every year. Surely it is here that we should be looking for improved efficiency. Such profligate waste is unaffordable.

Further, over one million kilometres of our river systems have been incised so that they are now flowing below their flood plains and are thus disconnected from them (SOE 2011, Ch. 2). In addition to improving soil organic carbon, we need to regenerate our wetlands and riparian zones, slowing the flow of water and reconnecting creeks and rivers with their flood plains. This will improve absorption of overflow for productive use in dry times and further recharge groundwater stores. Slowing down run-off will also limit the loss of nutrients from the soil.

Vegetation

Vegetation provides the link between soil health and water-use efficiency. On the one hand, it adds organic matter essential to improving the structure of the soil, enhancing water retention and further supporting ongoing plant growth. On the other hand, a protective cover of vegetation controls evaporation and soil loss through wind and water erosion (SOE 2011, p. 295).

Plant life diversity is important to support the ecosystem services that vegetation provides that are fundamental to human survival (SOE 2011, p. 268). A diversity of vegetative cover supports diverse microbial communities and healthy root ecologies, improving soil mineral health and facilitating effective nutrient transfer. Different root structures also access nutrients and moisture from different depths in the soil.

Trees can lower the water table and drawdown salinity from the soil surface. Trees also moderate temperature – limiting the effects of temperature extremes on surrounding groundcover. The restoration of functional vegetative cover also has the capacity to facilitate a cooler landmass by re-establishing the small water cycle to bring about a more even and regular distribution of precipitation (Kravcik *et al.* 2007).

Importantly, vegetation also provides the means to enable the drawdown of carbon dioxide via photosynthesis to be bio-sequestered into soil carbon sinks. We are aware of some innovative farmers who are sequestering up to 10 tonnes of carbon per hectare through their regenerative farming practices, improving soil organic carbon by over 200 per cent in just 10 years.

Soils for Life program

Our case studies provide some valuable examples of farming practices that have improved landscape regeneration. Many of our Soils for Life case study participants address soil health by actively managing the structural, mineral and biological elements. They do this through applying natural bio-fertilisers such as humus compost, compost teas, worm juice and biological fertilisers to stimulate soil microbes. Non-organic chemical inputs have been reduced or stopped. Minimal tillage cropping practices without the use of herbicides and pesticides are helping to minimise disturbance to the soil and protect microbial communities living within it. Stock are also used to contribute organic matter to improve soil structure and water retention.

Natural hydrological processes have been restored through the construction of swales or 'leaky' weirs to slow water flow to assist in healing erosion and recharging in-soil reservoirs. These case studies provide evidence of streams that now flow considerably longer and water being held in the landscape to be used by plants and animals – effectively extending the growing season. These practices have directly resulted in enhanced land resilience and productivity.

Each of our case study participants emphasises the importance of vegetation in their regenerative landscape management practices, especially in maintaining groundcover.

Grazing management, through various forms of time-controlled planned grazing, has been demonstrated to improve the diversity and extent of groundcover, especially of more resilient perennials, fertilise and improve the soil and enable increased carrying capacity on the landscape.

Some are using 'pasture cropping' techniques to direct drill crops into dormant native perennial grasses or under-sowing legumes or cover-cropping to protect the soil surface and return nutrients.

Others are experiencing improved agricultural production having provided shade and shelter for stock and crops through establishing shelter belts and wood lots. These farmers have reported that tree canopies have protected pastures increasing their production and reducing evaporation loss (SoilsForLife 2012, p. 29).

Each of these methodologies is not employed in isolation, as innovative farmers recognise that it is the art of managing the integration of soil, water and vegetation that is key to success. The Soils for Life program disseminates the results of these 19 case studies through workshops, field days and other engagement with the farming community and others engaged in food and fibre production. The full report, individual case studies, details of events, various resources and information are available and regularly updated on the website <www.soilsforlife.org.au>.

We are now planning to extend our case studies from 19 to 40 and continue to host field days across the various regions of our case study farmers and, where possible, in partnership with local community organisations such as community management agencies (CMAs) and national resource management groups (NRMs). The purpose of these demonstration days is to inform farmers and land managers of the options available to them and how best they can manage change.

Of the 130 000 plus farmers across Australia, some 1 per cent who have implemented innovative changes are reaping the benefits. Some 10 per cent are interested in investing in change, but the many perceived impediments to change persuade farmers that it is too risky. To date there has been too great a disconnection between research and farmers so we need to facilitate their re-engagement to learn from each other. We need a ground-up approach where increasingly farmers need to frame the questions and science comes up with answers (see Crawford 2013).

Soil advocacy

As the advocate for Soil Health I am charged with raising awareness of the importance of soil. This role involves the development of soil research priorities to complement existing efforts to develop a national soil research, development and extension strategy. My role is to help farmers to improve the health of their soil as this underpins not only biodiversity but also sustainable agricultural productivity across Australia (Prime Minister 2012).

As advocate I am working with scientists, farmers and business people. We need a collaborative approach and together I want us to ask fundamental questions relating to soil.

As a first step, I would like a clear definition of soil fertility and in that context how fertile are our soils and what exactly will each soil type grow? I would also like comprehensive data from a common database as to what parts of Australia are critical for the production of quality food and fibre and see science apply its best efforts to promote production in these areas. I would like to see a uniform national methodology for collecting, collating, analysing, distributing and storing information utilising uniform terminology.

Other important data needs include an effective measurement of soil carbon content to commercial standards and the completion of a comprehensive stocktake of our water resource: how much do we need, where is it, how much do we waste and how do we return and retain more of it in our soils?

The widespread adoption of regenerative landscape management is a strategic imperative for Australia's future wellbeing. I am looking to play a role with my team through the linking of a bottom-up Soils for Life, fixing the paddock approach with a top-down Soil Advocacy, fixing the research and policy approach.

Conclusion

A national, bipartisan policy framework that would underpin a policy to regenerate the health of the Australian landscape by restoring and maintaining an Australian landscape that is fit for purpose would:

- recognise soil, water and a diversity of vegetation as the nation's three key national, natural strategic assets, to be managed accordingly and in an integrated way
- reward farmers fairly, not just for their product but as primary carers of the agricultural landscape
- refocus science by asking of it the right questions and developing a common national information collection, collation, analysis and distribution chain
- re-engage urban Australia with its rural roots. This may well be the most important policy driver of them all
- recognise that to develop a proper landscape regeneration policy involves inputs not just from agriculture and environment, but education, health, regional development, indigenous, trade and national security and therefore needs to be coordinated at the highest federal/state/territory political levels
- note the potential of a properly managed landscape to draw down CO_2 and sequester carbon.

References

Chappell A, Webb NP, Butler HJ, Strong CL, McTainsh GH, Leys JF, Viscarra Rossel RA (2013) Soil organic carbon dust emission: an omitted global source of atmospheric CO_2. *Global Change Biology* **19**, 3238–3244.

Crawford J (2013) Discussion at the Jillamatong landscape rehydration in action during field day. Available at <http://www.soilsforlife.org.au/announcements/demo-day-landscape-rehydration-in-action> accessed 8 November 2013.

Kravcik M, Pokorny J, Kohutiar J, Kovac M, Toth E (2007) *Water for the Recovery of the Climate — A New Water Paradigm*. Available at <http://www.waterparadigm.org/download/Water_for_the_Recovery_of_the_Climate_A_New_Water_Paradigm.pdf> accessed 8 November 2013.

Prime Minister (2012) Address to the National Farmers' Federation National Congress, Canberra 23 October. Available at <http://www.nff.org.au/media-files/459.html> accessed 8 November 2013.

SOE (2011) 'Australia, State of the Environment Report'. Department of Sustainability, Environment, Water, Population and Communities, Canberra.

SoilsForLife (2012), *Innovations for Regenerative Landscape Management: Case Studies of Regenerative Landscape Management in Practice*. Outcomes Australia, Fairbairn, ACT.

11

Food, land and water: lessons from the Murray–Darling Basin

Rhondda Dickson

The story of the last 100 years of water management in the Murray–Darling Basin sets the scene for how we are placed today to meet the future challenges of climate change and increasing food demand.

This is a story of nation building and overcoming the huge challenges of developing agriculture in the driest continent with arguably the most variable water supply on Earth. It is a story of overcoming fierce interstate rivalries over water shares for the common good. It is a story of how governments and communities dealt with the significant environmental decline of the river system over the past 50 years.

And lastly it is a story of getting the right policy settings in place to underwrite a sustainable and productive future for all those who depend on these precious resources and landscapes. Australia has done this by setting a sustainable limit on water use, unleashing the innovation of farmers through the power of the market, and having flexible and adaptive systems that can respond to known and new challenges.

The Murray–Darling Basin contains 23 river valleys and Australia's two longest rivers, the Murray and the Darling. It has significant environmental values, and its

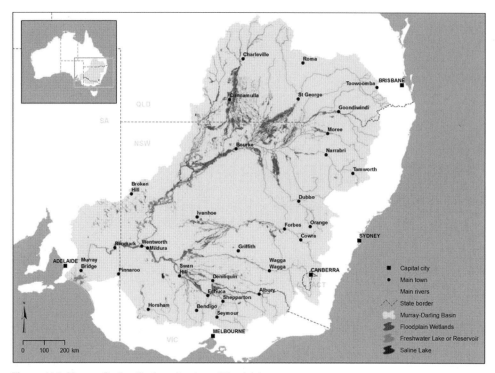

Figure 11.1: Murray–Darling Basin wetlands and floodplains.

slow rivers and flat landscape have created over 30 000 wetlands, of which 16 are Ramsar listed. It covers over one million square kilometres, or about 14 per cent of Australia (Figure 11.1).

Like most large river basins in the world, it has many demands on it. Millions of people depend on it. It is home to two million people and supplies drinking water to three million people, including people outside the Basin in Adelaide. It is the traditional home of over 40 indigenous nations who have the oldest continuous culture in the world.

It is often called Australia's food bowl – economically it is very important. In 2011–12 the gross value of irrigated production in the Basin was $6.7 billion, almost half of Australia's total irrigation production. This agricultural wealth is the result of nearly a century of development.

The focus in the early development of the Basin was on overcoming the challenges of the Australian climate, especially the boom and bust nature of the river system. Early colonial settlers were challenged by extreme floods and extreme drought.

Development could only take off in a big way, however, after the states settled their differences over the sharing of the River Murray and agreed on a long-term development program for the river.

Figure 11.2: The first River Murray Commission.

Rivalry between the states over ownership of the water along the Murray was intense for decades in the 1800s. In 1914 Victoria, New South Wales, South Australia and the Commonwealth governments reached a settlement on sharing the waters of the Murray that has largely endured to this day, the River Murray Waters Agreement (Figure 11.2).

The agreement was Australia's first trans-boundary water agreement. It presided over 70 years of river regulation, dam building and irrigation development (Figures 11.3 and 11.4).

As development and water use grew, however, problems began to emerge. Salinity was recognised as a problem in the 1970s. In 1981, the Mouth of the Murray closed for the first time in recorded history. In 1991, a world record algal bloom stretched 1000 km along the Barwon and Darling Rivers.

This environmental decline increased in parallel with the increasing water use. Between 1970 and 1994, water use in the southern Basin doubled.

To deal with the challenges Basin governments and communities began the long journey to establishing a sustainable limit on water use, and increasing the efficiency and productivity of irrigation industries.

Figure 11.3: Hume Dam.

Figure 11.4: Yarrawonga Weir during construction in 1936.

Governments needed to address over-allocation. The first step in doing so was to halt growth of water use. After some years of debate, governments agreed to cap diversions in 1995. The cap limited the growth and water course diversions levelled out.

The cap was always recognised as a holding pattern, however. It was not enough to address the continuing decline of the river system. In 2004, the first major national water policy, the National Water Initiative, was agreed. Under this national policy all governments committed to reduce over-allocation. The Living Murray Initiative was the first practical step in addressing over-allocation, by recovering 500 GL for the environment and committing to build a series of engineering projects that would deliver environmental water to important wetland areas along the Murray (Figure 11.5).

In 2007, in the depth of the millennium drought, the Australian Government announced the biggest water reform initiative in its history – a $10 billion package for investment in irrigation modernisation and environmental water recovery underwritten by reforms legislated in the *Water Act 2007*.

The Water Act established new whole of Basin governance for the Murray–Darling Basin and new institutions including an independent Murray–Darling Basin Authority (MDBA) to manage the river system in the national interest through a Basin Plan. The fundamental requirement for the Basin Plan was to establish a sustainable level of take.

In 2012 the Basin Plan became law and established a sustainable level of take for all rivers of the Basin. It required the recovery of a further 2750 GL for the environment by 2019.

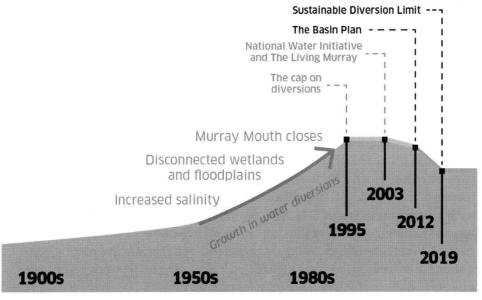

Figure 11.5: History of water diversions in the Murray–Darling Basin.

This level of recovery will not return the Basin to natural conditions – this was never the intention and realistically we cannot do this without removing all the dams and structures in the river. However, we can reduce the risks to the environment by ensuring that it is in a better condition so that it can weather droughts and floods and adapt to climate change.

Having a sustainable cap also enables markets to operate to maximise the productive potential of the available water. Under the National Water Initiative governments provided greater water security for irrigators through secure tradeable water entitlements. The Basin Plan provides the certainty of a sustainable limit and known rules. Combined they give irrigators the certainty to invest and innovate.

Under the National Water Initiative, governments introduced water trading to increase the efficiency of water use and improve productivity. During the drought the capacity to trade enabled the overall value of production to be maintained despite the dramatic reduction in water. Trading allows irrigators to be more flexible in their business decisions and change what they grow depending on the water available, such as choosing to grow irrigated crops in the high water years, but stick to horticulture in the dry years.

Separating water from land and allowing trade has also driven significant improvements in water use efficiency. With the ability to trade, the drought, rather than being just a threat, became a great driver of innovation. These innovations and improvements in water use efficiency have been supported by major government investment in modernising irrigation.

The investment by the Australian Government has been critical to support the transition to a sustainable limit. It has enabled the industry to be as efficient as possible through irrigation infrastructure efficiencies, modernising metering and allows for water purchase at market rates, which has allowed many enterprises to restructure the businesses.

Another important innovation of Australian water management is the water allocation system. Water licences were first defined in Australia in the *NSW Water Act 1912*. Over time a seasonally adjusted water allocation system evolved to help manage the pressures of a highly variable river system.

Water allocations in the Basin, in fact in much of Australia, are not a guaranteed volume of water that the user gets each year. Rather, the user is guaranteed a proportion of the available water in a given year and then allocated a volume of water each year based on how much water is available in that year. The water year begins in winter (May or June, depending on the state). After setting aside a minimum volume for things like critical human water needs, initial water allocations are based on the water that is currently in the dam and a worst case assessment of likely inflows through the year. This gives water users a minimal allocation at the beginning of the year, for example, Figure 11.6 shows a starting

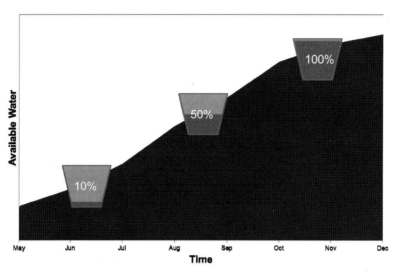

Figure 11.6: Seasonally adjusted allocations in a wet year.

allocation of 10 per cent. As the year progresses, and rain arrives to fill up the dam, allocations are gradually increased, potentially up to 100 per cent if it is a wet year.

If it is a dry year and very little rain arrives, allocations again start low and may stay low throughout the year. As these water entitlements are a share of the available resource, if water availability changes, the water allocations are able to adapt to an appropriate level. This ensures we never allocate more water than we have. This adaptable approach also inherently deals with a changing climate, by allocating lower volumes if the climate is drier (Figure 11.7).

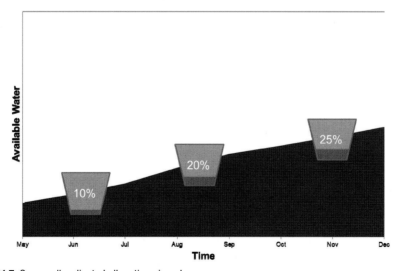

Figure 11.7: Seasonally adjusted allocations in a dry year.

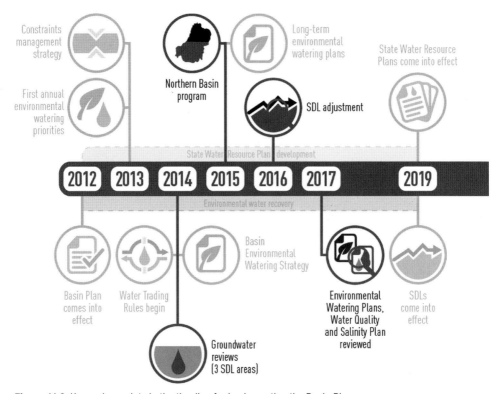

Figure 11.8: Key review points in the timeline for implementing the Basin Plan.

In addition to the year-by-year seasonal adjustments it is essential to have a flexible and adaptable planning framework, particularly in light of a changing climate. For example, the Basin Plan includes several opportunities for innovation and adaptation as it is rolled out, and the plan itself is reviewed every 10 years (Figure 11.8).

The differences in geography, rainfall, and water management in the northern Basin, compared to the south, means the northern Basin will require a different approach during the implementation of the Basin Plan. The MDBA has developed a Northern Basin Work Program to address some of these differences. The work program includes a review to assess new science about environmental water needs in the northern Basin.

When the Basin Plan was being drafted Basin water ministers asked that a mechanism to adjust the Sustainable Diversion Limits (SDL) be incorporated into the Plan. The mechanism adds flexibility to the Basin Plan by allowing the Murray–Darling Basin Authority to propose adjustments to the SDL in 2016 within ±5% of the SDL, that is ±540 GL. It works to adjust the Plan recovery amount by reducing recovery if equivalent environmental outcomes can be achieved with less water, and recovering additional water for the environment if this can be done with neutral social and economic impacts.

The MDBA's Constraints Management Strategy involves investigating the river management practices, physical structures and river height limits that restrict the effectiveness of environmental watering in the Basin. Modifying these practices has the potential to significantly improve and extend the benefits of environmental watering. The aim of the Strategy is to improve the environmental outcomes achievable beyond current operating conditions by allowing better use of environmental water while avoiding, managing or mitigating impacts to local communities and industries. This is a big opportunity to improve the way we run the river.

The Environmental Watering Plan aims to achieve the best possible environmental outcomes using the increased, but still finite, amount of water made available by the Basin Plan. Once established, it will be the first time environmental watering will be coordinated Basin-wide across state and territory borders. The plan is built on a principle of adaptive management, meaning the Plan will change and evolve over time to incorporate new knowledge, improved hydrological modelling, prevailing weather and climate conditions, previous outcomes, changing priorities and the requirements of different sites. Adaptive management also builds flexibility into planning. For example, heavy localised rainfall may provide managers with the opportunity at short notice to flood a river red gum forest using a minimal amount of additional water. The Environmental Watering Plan will also be reviewed in five years in light of experience.

The Basin Plan builds on the legacy of sustained reform of the previous decades where the Basin governments and communities have worked together to resolve differences, and learn to live with their environment. The result is a system that is able to manage the challenges of the future through three key features: a sustainable limit to water use; a strong industry, harnessing the market capacity and innovation of Australian farmers; and a flexible and adaptable framework.

12

Balancing water use for food and the environment: looking to the North based on lessons from the South

Gary J. Jones

Setting the scene

For more than a hundred years, irrigated agriculture has been a major driver of economic and social development in rural Australia, and in the Murray–Darling Basin in particular. But with that has come very significant environmental damage. This has seen not only deleterious impacts on native plants and animals but it has, in places, threatened the viability of the irrigation industry itself, not least due to soil salination and erosion.

Thankfully, a single positive outcome of this unsustainable development pathway has been our ability as a society to recognise problems when and where they emerged, and to implement appropriate policy and management responses (albeit, at times, slowly and reluctantly).

Well-known examples in the Murray–Darling Basin include the Salinity and Drainage Strategy in the 1980s, the Cap on surface water extractions in the 1990s and, most recently, the Living Murray and Basin Plan agreements to recover water for the environment.

Nevertheless, these actions have always been undertaken in retrospect, based on hindsight of the problems agriculture has caused for the environment and for itself, rather than being informed by foresight and truly good planning from the outset for a sustainable irrigation industry.

As Australia looks increasingly to its tropical northern lands as a prospective food-bowl for Asia we should reflect on two important questions:

1   Have we gained sufficient knowledge and wisdom from a century of unsustainable irrigation practices in southern Australia to do things differently in the future?
2   Is northern Australia really the agricultural utopia that some in the community argue, and do the potential rewards justify the risks to our largely pristine and biodiverse tropical river basins?

Understanding the causes of environmental damage arising from agriculture

Landscape and catchment impacts

The sustainable management of land and soils has always been a fundamental challenge for Australia and, unfortunately, it is something we were poor at for a very long time. The combination of unconstrained clearing of native vegetation, particularly in hilly terrains (Figure 12.1), combined with the erosive power of high-intensity Australian rainfall led to massive soil erosion problems in many farming areas.

This led to a double-whammy outcome where (i) farmers lost fertile topsoils and damaged their lands through gullying (Figure 12.2A), and (ii) the river environment downstream of these agricultural areas suffered as farm soils and sands were washed away during heavy rainfall, muddying river waters and smothering stream-bed habitats essential for the survival of riverine biota (Figure 12.2B).

Along with soil and fine sediment, fertilisers and pesticides may be transported from farms into rivers, wetlands and coastal water, especially when they are applied poorly or in excess of crop needs. These too have a negative impact on receiving waters – either by stimulating unwanted algal blooms or by being toxic to native animals in the river system.

The positive news is that thanks to a combination of good scientific and agronomic research over the past two to three decades, combined with extensive

Figure 12.1: Hills denuded of vegetation. (Photo: John Coppi/ScienceImage.)

on-ground trials by landholders (funded and carried out through various programs such as Landcare and various regional National Resource Management (NRM) groups), we now have a reasonably good handle on the most appropriate agricultural practices for stopping erosion and retaining soils on-farm, and for minimising fertiliser and pesticide run-off. For example, in the Great Barrier Reef coastal catchments, restoration and prevention programs are in now place to improve farming practices which will minimise soil and fertiliser wash-off.

Another major problem in Australia agriculture has been soil salination. This has two different settings and causes, arising separately on dry-land and irrigation farms. In both cases, mobilised salt can travel from farms back into adjacent rivers – by natural run-off processes or via irrigation drains – causing river salinity problems tens or hundreds of kilometres downstream. This was the situation in which the people of Adelaide found themselves in the 1960s and 1970s, eventually necessitating huge engineering interventions[1] to stop the salt from reaching the River Murray and Adelaide's water supplies that are drawn from it.

From a river ecosystem perspective, the biggest impact arising from the development of irrigated agriculture in southern Australia, indeed throughout the world, has been the building of large dams on rivers to store and distribute water (Figure 12.3A).

Dams on rivers have two major types of ecological impact. First, there are physical impacts – they are a barrier to the necessary upstream and downstream

Figure 12.2: A. Hillslope gully and sheet erosion, northern QLD. (Photo: Willem van Aken/ScienceImage.)
B. Murrumbidgee River outside Murrumbateman, NSW. (Photo: Nick Pitsas/ScienceImage.)

A.

B.

Figure 12.3: A. Burrinjuck Dam, Murrumbidgee River. (Photo: ©Commonwealth of Australia (Murray–Darling Basin Authority), Photographer Irene Dowdy, 2009.) **B.** Off-river storage on a northern MDB cotton farm. (Photo: ©Commonwealth of Australia (Murray–Darling Basin Authority), photographer Arthur Mostead, 2010.)

movement of aquatic animals, especially fish. Second, there are hydrological impacts – by capturing water for irrigation, dams reduce downstream flow volumes and velocity, and also change the timing and pattern of flows.

The hydrological changes can be highly detrimental to river biota, especially to the plants and animals living or breeding on the river's flood plain. They rely on regular flooding to sustain their growth or to stimulate seed germination or animal breeding. Small to medium-sized floods, occurring every year or so, are of particular ecological benefit and it is these that are most reduced by dams (large floods pass through a river system more or less unaffected).

The dampening of the natural variability in downstream flows by dams also impacts on fish in the river channel, which rely on certain flow velocities or water depths, for example as a cue for migration. Water released from large dams is also often much colder than that naturally flowing in a river and this may also impact negatively on fish breeding.

On-farm dams – large and small – can also cause serious eco-hydrological problems, even if individually they are much, much smaller than on-river dams. When there are many in a catchment, across many farms, their combined impact on run-off and river flows can be significant (Nathan and Lowe 2012).

Further, in the so-called unregulated reaches of the northern Murray–Darling Basin where there are no large dams on rivers, large on-farm dams harvest river waters during flood times for later use on crops, mostly cotton (these are often known as 'ring-tanks' because of the way they are constructed) (Figure 12.3B). While this might sound harmless enough, perhaps even beneficial, the combined impact of large ring-tanks on downstream river flows and biota can be serious.

Smaller weirs are built on rivers to provide a local head of water to allow gravity supply of water for irrigation (and for some towns). We learned early on that fish cannot move up and down the river to feed or breed if there are weirs blocking their path. Some weirs are removable or have gates that can be fully opened at certain times of the year to allow fish to move past. In other weirs, fish ladders were built to allow fish to move past them. Unfortunately our original designs were taken from Europe and were based on the behaviour of salmon. Salmon are fish that jump but our sluggish Australian fish are not much good when it comes to jumping!

After some good local research in the 1980s and 1990s, we realised that ladders could be designed to better suit Australian fish (Figure 12.4), and even fish lifts have been built where fish can swim in at the bottom, get lifted up in a cage and swim out upstream at the top.

Finally, one other thing that that has been learned is that clearing vegetation right down to the water line of the river is not a good idea. The stream-side or riparian vegetation plays many key roles in maintaining a healthy river. It filters out (some but not all) nutrients running off the land before they reach the stream,

Figure 12.4: Modern Australian fish ladder – Torrumbarry Weir, River Murray. (Photo: Sarah Cartwright, CRCFE.)

as well as stabilising river banks, shading smaller streams and otherwise providing habitat for animals, aquatic and terrestrial.

Now that we properly understand the importance of river riparian corridors and the impacts of clearing, much restoration work has been undertaken. This includes physical works to reshape river banks (where badly eroded) and the re-establishment of endemic vegetation (and the removal of invasive, exotic species such as Willows where required).

One other catchment-scale impact, while largely out of sight, that should not be ignored is the (unsustainable) use of groundwater. Where pumping by farmers exceeds the rate of replenishment,[2] the groundwater level decreases, making pumping more expensive or even impossible in extreme cases. At the same time, unsustainable pumping can negatively affect so-called 'groundwater-dependent ecosystems' (Murray *et al.* 2003). These could include certain types of wetlands where water supply from below is important (other wetlands rely on surface run-off only), some woodlands and the mound springs of the Great Artesian Basin.

Local or habitat impacts

The range of local impacts of agriculture on river and flood plain habitats – the places where plants and animals live, feed and reproduce – is broad. Many are directly linked to the catchment-scale impacts outlined above, while there are others that arise due to distinctly local factors. Habitat degradation linked to agricultural practices in a river's catchment includes:

- sand smothering of riverbed habitats – impacts particularly on invertebrates that live and feed on the bottom of streams, and which are also the food source for many fish and other animals (e.g. platypus) (Figure 12.2B)
- fine sediment run-off – makes water more turbid with lower penetration of sunlight into the water. In turn, this affects the ability of plants to grow below the water's surface. These submerged plants are an important part of healthy river ecosystem
- fertiliser run-off – stimulates the growth of nuisance, filamentous algae that grow on submerged logs and rocks – these crowd out the formation of natural microbial biofilms that are a more palatable food source for river animals
- changed local water depth, flow velocity or water temperature caused by upstream dams – may impact directly on the local habitat suitability for many animals including fish, turtles and mussels.

Habitat degradation that is not linked to upstream catchment condition, but arises due to local farming impacts, includes:

- edge habitat destruction – caused by cattle given direct access to the river for watering. Many riparian restoration programs across Australia now fund farmers to fence off their lands from the river and to provide alternative watering points. Cattle defecating in streams under these circumstances also create local pollution problems as well as further downstream (including potential human health problems from drinking water which is contaminated by animal intestinal parasites such as *Cryptosporidium* and *Giardia*).
- levees and block-banks on the flood plains – farmers with lands adjacent to rivers may construct levees to divert minor floodwaters away from, or towards, certain parts of their property. While the local ecological effects of this may be minimal, there are situations where such works cause the drying out of wetlands or woodlands further along the flood plain.

Lessons learned

The upside of this wide range of impacts that has arisen through the development of irrigated agriculture is that, as scientists, managers, farmers and concerned citizens, we have learned a great deal about how *not* to go about developing and maintaining a large, productive agricultural system! For any new irrigation development, including any proposed for northern Australia, we can reasonably claim knowledge of the impacts of past agricultural practices and that we have learned ways of carrying out irrigated agriculture far more wisely.

To summarise, we have learned that sustainable irrigated agriculture *should* include the following catchment and farm-scale practices:

- clearing land sufficient to grow crops and no more, retaining as much native vegetation on-farm as possible
- protecting vegetation along riparian zones, including fencing where necessary, and on steeper slopes or other areas of higher erosion risk
- adopting modern tillage and other agronomic practices that maximise water and soil retention on farm (and that enhance soil fertility)
- applying water sufficient to meet crop needs and no more, using modern high-efficiency irrigation delivery systems (such as micro-irrigation, pressurised supply, etc.)
- controlling the application of fertilisers and pesticides, at the lowest practical levels, and retaining any drainage waters on farm (unless otherwise proven safe to discharge)
- avoiding or minimising the need for dams, especially large dams. If dams must be built (and, to be clear, this is not desirable) the combined storage volume of dams on a river system should be *much* less than the mean annual run-off upstream of where the dam is to be built. (Note: in the Murray–Darling Basin the combined dam storage volume is 1.5 times mean annual run-off, hence the huge magnitude of their ecological impacts.)
- implementing regulatory controls on the construction of farm dams and flood plain banks and levees
- adopting the combined and sustainable use of surface water and groundwater – for example, between wet and dry seasons. This should be optimised to maximise supply reliability and to minimise impacts on all water-dependent ecosystems
- applying ecologically defined limits on the total amount of water that can be withdrawn for irrigation from a river or groundwater system in any season/ year.

There are many other sustainable practices that should be adopted – this list is meant to be illustrative rather than inclusive. The key message is that ecologically sustainable irrigated agriculture is technically feasible, if we have the intelligence and the conviction to implement it fully.

Whether or not politicians have the motivation or the will to fully implement the required approaches and practices is another question. Ecologically sustainable irrigation comes at a cost – in increased system development and operating costs, in reduced area of agricultural farmlands or water available for production, or both. But the costs of environmental degradation itself are economically real – not just unforeseeable externalities – although they may not be observed for years or decades, certainly well beyond the lifetime of most politicians.

Herein may lie the dilemma for irrigated agriculture in northern Australia.

Is there a case for major expansion of irrigation in northern Australia?

Having argued that ecologically sustainable irrigated agriculture is, at least, technically feasible (if politically unrealistic!) the remaining fundamental question is this – is Australia's tropical north the potential agricultural utopia it is claimed by many to be?

The northern regions in question stretch from Broome in Western Australia to Cairns in north Queensland (Figure 12.5). There is already rangeland grazing through the region, and some irrigated agriculture, for example in Western Australia (Ord River), Queensland (Flinders and Gilbert Rivers) and the Northern Territory (Daly River).

The idea that northern Australia should be a good place to develop agriculture has been pervasive for many decades, in spite of well-documented arguments to

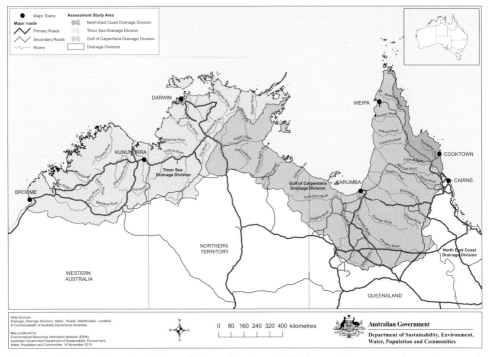

Figure 12.5: Northern Australian drainage divisions. (Department of Environment – <www.environment.gov.au>.)

Table 12.1: Key findings in relation to developing water and land resources in northern Australia (CSIRO 2009)

• There are critical gaps in our knowledge and data sources, and in our understanding of Indigenous knowledge.
• Despite high rainfall, the north is seasonally water limited.
• The ability to capture and store surface water for consumptive use is constrained by climate and topography.
• The development of groundwater resources provides the best prospect to support new consumptive uses of water.
• There are approximately 600 GL of groundwater *(only)** potentially available across northern Australia that could support new consumptive uses.

* My addition for emphasis.

the contrary (for example, Davidson 1965). The region has almost 50 per cent of Australia's total surface water run-off, a big contrast to the run-off in the Murray–Darling Basin of only 6 per cent, and where there is already billions of dollars-worth of agricultural productivity. It is captivating in a food-hungry world to think we might harness those northern water resources and have more agriculture and more production. Naturally, it is not quite that simple, or development would have proceeded long ago (notwithstanding the initial faltering attempts for the Ord Irrigation Scheme).

A relatively recent set of scientific, economic and cultural studies into the development prospects of the north was reported by the Northern Australia Land and Water Taskforce (CSIRO 2009) (Table 12.1).

One key point in Table 12.1 is that, despite the high average rainfall, the north is seasonally water limited. This is not unusual in the tropical latitudes, being also the case in the Philippines, Indonesia and parts of southern India. In all of those countries, a higher than average annual rainfall does not guarantee a year-round water supply.

A further important point noted by the Northern Australian Land and Water Task Force (CSIRO 2009) is that the most useful rainfall tends to be away from areas where the soils are suitable for irrigation. Also, there are not very many good locations for big dam sites in the north (notwithstanding the major ecological impacts of large dams already discussed). Although Lake Argyle, the second largest dam in Australia, has been built on the Ord River in northern Western Australia, there are not many dam sites like that and, as already noted, it is to be hoped we are wise enough now to know that building such large dams is not the way of the future.

CSIRO and others acknowledge, however, that there is potential for some new irrigated agricultural development in northern Australia, largely based on complementary use of surface water and groundwater, between wet season and dry season. But the extent of this is nowhere near that suggested by some.

Prior to the 2013 federal election, the Liberal–National Opposition (now Government) released *Coalition 2030 Vision for Developing Northern Australia*, their vision for developing northern Australia. That policy document stated that Australia's agricultural production could be doubled by 2030 with such northern development.

However, CSIRO and the Northern Australian Task Force had already concluded that only about 600 GL of groundwater is available to sustainably support production, which is a very small amount when compared to the greater than 10 000 GL of water typically used by irrigated agriculture across Australia (in non-drought years). On an area basis this would mean an additional 40 000– 60 000 hectares of irrigated farmland,[3] which is minor compared to that already in Australia of 1.8 million hectares.

It was concluded that achievable extra irrigated agricultural production in northern Australia is likely to be only around 3 to 4 per cent – a long way from the doubling proposed by the (now) Federal Government.

The Flinders and Gilbert catchments in the elbow of the Gulf of Carpentaria feed significant river systems and, during 2013, the Queensland Government released about 94 GL of that surface water for new irrigation developments there. Again, this is not very much water in comparison with the volumes in use for irrigation in southern regions, although it is evidence that some new agricultural development is underway in the north.

From an ecological perspective, there is a pertinent warning by Georges *et al.* (2002) based on work in the Daly River that overuse of groundwater is likely to interfere with habitat for at least some of the north's unique freshwater species, especially turtles and other amphibians.

One other important aspect that must be understood and respected in the north, particularly around the Gulf of Carpentaria, is that successful coastal commercial and recreational fisheries for Barramundi and Banana Prawns are highly dependent on the amounts of water that flow down the rivers and out to sea. If too much water is taken out for agriculture upstream, the coastal fisheries are likely to suffer. This point was made recently by Michael Douglas who leads the Tropical Rivers and Coastal Knowledge (TRaCK) research hub based in Darwin (TRaCK 2012):

There is a perception that wet season flows in the north are 'wasted', but our research has shown a direct correlation between river flows and the commercial and recreational catches of coastal fish such as barramundi, king threadfin, and prawns. ... Large floods that spill over the banks allow fish to move onto the flood plains to feed then move back out to the river as the flood plains dry. ... This means that much of the meat on barramundi in the upper reaches of river systems may have been 'grown' on the flood plains, potentially many months before and hundreds of kilometres downstream, and similarly, freshwater flows into estuaries play a significant role in determining the numbers and sizes of fish that live there. ... These floods are also necessary to connect the flood plains to the river and allow movement throughout the entire river system, while maintaining dry season flows may be essential for the life cycles of important species such as barramundi and freshwater prawns.

Conclusion

Australia already produces enough food to feed 60 million people, of which we export almost two-thirds. Australia is number 10 among net food exporting countries in the world. We have more than enough food to feed ourselves for many decades, if not centuries, to come.

We have learnt much from the environmental mistakes we have made in developing profitable agricultural production in southern Australia over the past century or more. Agricultural productivity and efficiency are continuously improving. We are managing farming lands better while simultaneously recovering, where necessary, water for the environment to remedy past impacts on riverine ecosystems.

As and where needed, and notwithstanding some regional scientific and indigenous knowledge that still requires better gathering and valuing, we have learned enough from our past mistakes in the south to expand irrigated agriculture in the north in a resource-sustainable way while avoiding significant ecological damage.

The tough questions remaining, however, are these:

1   Do we have the political will to fund and develop irrigation in the north in a proper, sustainable manner?
2   And even if so, are the apparently modest economic gains from irrigated agricultural expansion in northern Australia worth the risk of losing the cultural, tourism and other benefits of such pristine and biodiverse lands and river systems?

As has been noted by others, there is probably a far better case for driving greater irrigation efficiencies in southern Australia than for further developing the north. Here there are already well established food production and market-delivery systems, and the scale of agricultural production is an order of magnitude greater than anything that appears viable in northern Australia. And to be frank, in the south the ecological damage has already been done and is now, if slowly, being repaired.

Acknowledgements

Ann Milligan is thanked for editorial support.

Endnotes

1   These consisted of a network of salt interception bores and discharge/evaporation basins. See: <http://www.mdba.gov.au/annualreports/2009–10/chapter3–6.html>.
2   Replenishment of groundwaters occurs by local rainfall infiltration or transport from further afield via connected aquifers.
3   Currently there are about 34 000 hectares of irrigated lands in northern Australia (CSIRO 2009).

References

CSIRO (2009) 'Northern Australia Land and Water Science Review 2009'. <http://www.csiro.au/Organisation-Structure/Flagships/Sustainable-Agriculture-Flagship/Northern-Australia-Sustainable-Development/Science-review-key-findings.aspx>

Davidson BR (1965) *The Northern Myth: A Study of the Physical and Economic Limits to Agricultural and Pastoral Development in Tropical Australia*. Melbourne University Press, Melbourne.

Georges A, Webster I, Guarino E, Thoms M, Jolly P, Doody S (2002) 'Modelling Dry Season Flows and Predicting the Impact of Water Extraction on a Flag Ship Species'. Final Report for Project ID 23045. CRC for Freshwater Ecology, University of Canberra, Canberra.

Murray BR, Zeppel MJB, Hose GC, Eamus D (2003) Groundwater-dependent ecosystems in Australia: It's more than just water for rivers. *Ecological Management & Restoration* 4(2), 110–113.

Nathan R, Lowe L (2012) The hydrologic impacts of farm dams. *Australian Journal of Water Resources* 16(1), 75–83.

TRaCK (2012) New research casts doubt on northern food bowl. Tropical Rivers and Coastal Knowledge research hub, Media release May 2012. <http://www.track.org.au/news/2012/new-research-casts-doubt-northern-food-bowl>

13

Human health: bottom-line integrator of impacts of the population, resources and climate change

Anthony J. McMichael

How Australia responds to the conjoint pressures from population, resource depletion and degradation and climate change bears greatly on our environmental and social future. Much of this ancient continent is arid, with vulnerable soils, extensive coastline and coastal ecosystems, frequent weather extremes, restless regional neighbours, and skewed distributions of freshwater flows and human population settlement. Yet public discussion of the management of population numbers and flows and of resource stocks has been consistently superficial and fragmented; likewise, discussion of whether and how to constrain carbon emissions has focused on protecting economic growth, jobs and commercial competitiveness, *not* on protecting the planet and future generations.

Unabated, climate change threatens the ecological and social foundations of *population* health and survival. In addition to the well-recognised risks from increases in extreme weather events, climate change will damage population health by disrupting eco-social systems and thereby influencing food yields, freshwater supplies, infectious disease patterns, and refugee flows and the many attendant health risks. Those health impacts will be unevenly shared around the country. Meanwhile, population and climatic pressures increasingly threaten Asia–Pacific regional food security, river flows and habitable land – and will increase migration and displacement.

Mainstream climate change science has paid relatively little attention to the population dimension. At any level of per capita emissions, the more people the faster and greater the change in climate. To live sustainably within natural limits we must jettison crudely formulated Big Australia pretensions and reduce our overly-large per-person ecological footprint. We must also address the discomforting reality that 'Lucky' Australia has moral responsibility to accept, within a regionally agreed policy, a fair share of the region's climatically environmentally dispossessed – our emissions affect the *global* climate. True sustainability cannot be built on a national fortress-like mentality, the outdated legacy of Europe's Treaty of Westphalia (agreed in 1648) that spawned a chessboard world of self-protecting and economically self-interested sovereign states. That outdated policy mindset must evolve towards an historically unprecedented international sharing, a harnessing of 'The Better Angels of Our Nature' to living in a *global commons* (Pinker 2011).

The determinants of population health

In this chapter I explore some of the past, present and future connections between population, resources and climate change in Australia, and the risks to human wellbeing health and survival. I draw on several aspects of history, and pose some questions about the impacts of climate change on our regional neighbours and what that portends for Australia.

First, though, I will make an obligatory comment about the nature, determinants and significance of population health. The impacts of environmental degradation and disruption on human health and survival are too often viewed as regrettable *collateral damage*, a blemish on the margin. This constricted view reflects the misleading, neoliberal, assumption that health is primarily determined by personal behaviours, genes and access to health care. Yet at the more important *population* level, and over the longer term, the determinants of population health lie with nature's life-supporting systems.

Figure 13.1 summarises the three main types of paths by which climate change influences population health. Path 1 is very familiar. Paths 2 and 3 are qualitatively different; they entail more complex causal webs that involve disruption or depletion of nature's biophysical and ecosystem stocks and flows and these then impinge on the foundations of population health.

Many influences of climate change are already evident in Australia, or probably happening but difficult to discern, or anticipated in the future. They include:

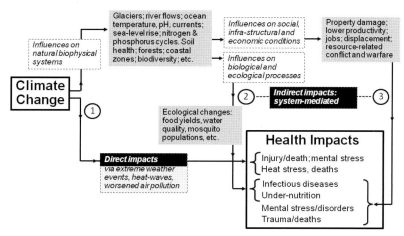

Figure 13.1: The main types of pathways by which climate change influences the health of human populations and communities.

- Already apparent: prior risks amplified by climate change
  - Uptrend in average annual number of heat-days → deaths, hospitalisations
  - Increasing number/severity bushfires → injury/death, respiratory hazard, mental health sequelae.
- Current probable health impacts: but not clearly identified/identifiable
  - Rise in food-borne diarrhoeal disease
  - Altered air quality: ozone formation, aeroallergens
  - Thermal stress in outdoor workers: behaviour, injury, organ damage; reduced productivity
  - Mental health impacts, particularly in some (drying) *rural* regions.
- Anticipated future health impacts
  - Extreme weather events: injuries, deaths, infectious disease, depression
  - Water shortages: food yields, hygiene, recreation
  - Mosquito-borne infections: dengue, Ross River virus, Barmah Forest virus, Japanese encephalitis, etc.

Many of these relationships are not tractable to conventional quantitative epidemiological research. Mostly, climate change and its resultant environmental changes act as a risk multiplier – or in some instances as a risk divider. In a Plus 4°C world, for example, parts of northern Australia could become too hot for the mosquito vectors that spread dengue or Ross River virus.

The direct health impacts of heat extremes are measurable via straightforward quantitative information, and the causal connections are fairly obvious. Such was the case with the extreme heatwave of summer 2009 that afflicted Melbourne, including three days of around 43–44°C. When temperatures soared above 40°C the number of ambulance callouts increased by up to twentyfold. Death rates tripled (Department of Health and Human Services (Victoria) 2009).

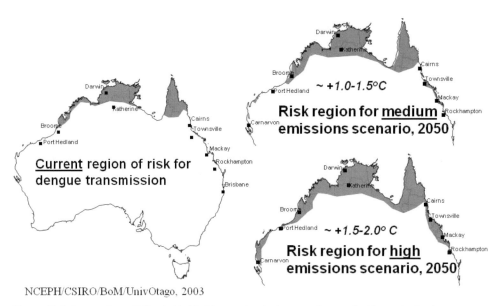

NCEPH/CSIRO/BoM/UnivOtago, 2003

Figure 13.2: Dengue fever: Estimated habitable zone for *Aedes aegypti* mosquito, the vector for dengue, under alternative CSIRO climate-change scenarios for 2050.

We have recently completed a different type of study on heat and health. The background temperature in Australia has been rising since the 1960s, during which there has been a progressive redistribution of the total yearly deaths between seasons – more now occurring in the warmer months, fewer in the colder months (Bennett *et al.* 2013). This is *suggestive* of a heat effect, especially since we allowed for time-trends in age distribution, and confirmed that this seasonal redistribution occurred for the major cause of death: heart and blood vessel diseases.

Path 2 in Figure 13.1 entails very different examples. Figure 13.2 summarises the modelled forecasts of potential dengue fever under climate change conditions that we produced for the Commonwealth Department of Health early last decade (McMichael *et al.* 2003). Modelling techniques then were rather rudimentary, and we are now constructing an ecologically based model based on field experiments and mosquito population data to incorporate the complexities of the component relationships in this transmission system.

This type of study cannot provide simplistic estimates of single-factor 'relative risks', such as the tenfold increase in risk of heart attack in heavy smokers. The measures, or qualitative summaries, need to accommodate more provisionality and uncertainty.

A second Path 2 example examines how warming is likely to cause the regional habitat of skipjack tuna in the Pacific to shift eastwards during this century. Skipjack tuna is a prime component of a healthy diet – protein and omega-3 fatty acids – and tuna-fishing is also the base of much local employment and of export earnings for many small Pacific Island states. Political strife may occur too, since relative sizes of the catch in each of the island-specific territorial waters will change. If regional negotiations *cannot* re-balance the fish-harvest equation, jobless people will be displaced from some islands and migration and refugee flows will increase ... and Australia would be a prime preferred destination for many. Inevitably assorted risks to both physical and mental health will afflict many of the displaced people.

There are five types of climate-related health impacts that I think are now evident, from the synthesis of multiple independent observations in various regions of the world. They are:

1   recent uptrend in adverse health impacts from cyclones, storms, bushfires, flooding
2   increasing annual numbers of deaths from heatwaves in a range of countries
3   shifts in range and seasonality of some climate-sensitive infectious diseases (and their vectors)
4   contribution to declines in food yields in some regions, with particular risk of malnutrition-related child development
5   adverse mental health consequences in various rural communities affected by drying.

The fourth point, relating to food yields, refers to a widespread and growing risk to human health. The combination of population growth, rising consumer expectations, a demand for more meat (especially grain-fed red ruminant meat), soil degradation, dwindling irrigation water supplies, biodiversity losses such as depletion of pollinators, all capped off by climate change, raises serious concerns about our capacity, as Earthlings, to feed 9.6 billion people by 2050. There is talk of a Second Green Revolution, and there may indeed be some genetically modified rabbits to be pulled out of hats. But, currently, most trends are not encouraging. Indeed, arable land is becoming a prized international asset for governments and large investors, many of whom sense that food shortages are on the horizon.

Australia's soils should be a great and crucial asset for sustainable living in future, including judicious regional exports. More immediately, we must sustain that arable and pastoral resource base to ensure its future capacity and resilience under climatic stress to feed Australians a healthy, though calorically more modest, diet. When food prices rise, on top of existing income inequalities, the nutritional health of poor families suffers, with lifelong adverse health consequences.

Our agricultural system generates much CO_2 along with two other potent greenhouse gases: nitrous oxide emissions primarily from fertilisers, and

Figure 13.3: Australia's Big Methane (burp) emitters. Ruminant mammals pre-digest the cellulose in grasses and other plants, which generates 'enteric' methane – a potent greenhouse gas – in large volumes.

voluminous methane from ruminant livestock, mostly cattle and sheep (Figure 13.3), whose fore-stomachs pre-digest cellulose in grass. The rising demand for beef in developing countries, some of it exported from Australia, thus puts further pressure on the climate system (and on Australia's pastoral resources and infrastructure). Indeed, worldwide, the livestock sector accounts for around 35 per cent of greenhouse gas emissions.

Historically, food shortages and famines due to natural climatic trends and fluctuations have been the great killer of populations and leveller of civilisations over thousands of years. In many cases, the local population had grown, during good times, well beyond the point of resilience to climate-related changes in regional carrying capacity.

Historical examples of food stresses and crises due to changes in regional climates abound, such as:

- the Sumerians around 4500 years ago, as regional drying in Southern Mesopotamia – occurring on top of already damaged soils – set in
- the decline of the Classic Maya in Central America in the 8th to 11th centuries CE, when an unusual sequence of droughts occurred
- the dissolution, around 1100CE, of the Tiwanaku civilisation, high on the altiplano adjoining Lake Titicaca at today's border between Peru and Bolivia, and dependent on glacier meltwater
- the uprising of the Chinese populace in the climatically dire first four decades of the 17th century, when starvation drove the peasantry and disaffected military to mobilise, invade Peking and, in 1644, storm the Forbidden City – causing the last of the Ming Emperors to hang himself
- and currently: Oxfam considers that the flight of starving Somalis into adjoining Kenya is the 'first humanitarian crisis due to climate change'. This

local famine accords with recent scientific reports that the recent warming of the western Indian Ocean has changed its surface current circulation. This has altered the usual pattern of regional moisture-bearing westward winds, the local monsoon system that provides most of the rainfall in the Horn of Africa. Failure of that rain has caused drought and harvest failure.

Consider the first two examples in more detail.

Sumer was a pioneering early farming society in southern Mesopotamia (approximately today's central-southern Iraq). Its experience shows well the interplay between population size, resource depletion, climate change and human impacts. During the period 6000–4500 years ago, the combination of population growth, deforestation, extended irrigation, soil salination and increasing urban demands for food led gradually to a seriously stressed food system. Then, in the following centuries the monsoon system shifted, regional drying emerged, food security worsened, hunger spread and local village communities built fortifications to protect their food stores from raiding neighbours.

Subsequently, the regional drying spread north and severely affected the expansionist Akkadian-Mesopotamian empire. Crops failed and hunger followed. Seeking water and pasture, many northern pastoralists moved down the Euphrates river valley with their goats and sheep. Conflict with Sumerian farmers broke out when these free-range herbivores found new fields of cultivated food to dine on. Despite wall-building by local authorities to 'Stop the Goats', the numbers of southwards-displaced persons were too great to resist. As the Akkadian troubles spread into southern Mesopotamia, conflict, crowding and crisis ensued and this now conjoint civilisation weakened and collapsed around 4100 years ago (deMenocal 2001).

Three thousand years on, there occurred the demise of the Classic Mayan civilisation in Mesoamerica, a topic that has long generated speculation and debate. That civilisation spanned what is now south-east Mexico, the Yucatan Peninsula and parts of Central America. It flourished during the first seven centuries CE, and then went into decline. New insight has come from remarkable recent gains in reconstructing past regional climates, and especially from two independent high-resolution proxy methods of estimating monthly rainfall over the past 2000 years – by micro-stratum chemical analysis of coastal sediment cores and of a stalagmite from the Central Mayan Lowlands (Kennett *et al.* 2012).

Over the course of three centuries, from around 750 to 1050 CE a succession of serious and prolonged droughts, previously inferred by archaeologists, occurred. They undermined food yields and social cohesion in the Maya; starvation and local conflict (evident in skeletal remains), social unrest and ultimately violent uprisings unseated the rulers of many of the great centres. This decline process was spread over a couple of centuries.

There are some warning signs in these and many other historical experiences for Australia. Meanwhile, there is a much longer human population and climate narrative potentially available in Australia. Yet there has been little discussion and information about the demographic, social and health experiences of Australian Aboriginal populations under climate change conditions over the past 40 000-plus years. But new information is now accruing, and glimpses of the distant past are becoming possible.

There have been great swings in Australia's climate over that long period, reaching back to the middle of the last glaciation around 50 000 years ago. Of particular interest is the dryland Aboriginal population's experience during the Last Glacial Maximum (i.e. maximum *cold*), around 20 000 years ago, and subsequently during the wobbly warming through to around 12 000 years ago as the Pleistocene came towards its end and Australia entered the warmer and climatically *relatively* more stable Holocene.

That dryland hunter-gatherer experience illustrates the interconnections between climate, the environmental resource base and population size – although population size was primarily an outcome rather than an input. During the Last Glacial Maximum, temperatures fell as low as 10°C below today's temperature, accompanied by declining rainfall and widening aridity. Lakes dried up, deserts expanded, and various animals went extinct. Around three-quarters of Australia's land area was abandoned as people sought refuge in well-watered riverine sites.

Archaeological data from 908 dryland sites, representing around two-thirds of Australia's land surface, indicate a saw-tooth pattern of population growth and decline, on a 1000–3000-year frequency (Figure 13.4). Major step-wise increases

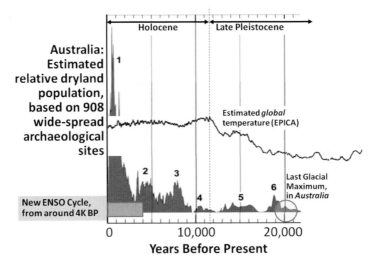

Figure 13.4: Fluctuations in Australia's dryland Aboriginal population since the end of the last ice age (glaciation). Actual numbers could not be estimated, but were estimated in proportion to archaeological evidence of activity and physical range. (Adapted from Smith *et al.* 2008.)

are evident around 19 000, 8000 and 1500 years ago – shown here by numbered peaks 6, 3 and 1 (Smith *et al.* 2008). The estimated global temperature, from polar ice-core analysis, is also shown.

A major population decline occurred around 17 500 years ago (following peak 6). The combination of incipient warming, persistent aridity, the need to reorganise land use and population dispersal from the refuge sites may have disrupted cultural practices and living conditions (Williams *et al.* 2013). The archaeological evidence indicates that during this radical change in climatic conditions big adaptive changes in settlement and subsistence patterns occurred, including changes in hunting practices and types of food eaten.

Changes in the relative size of the dryland population size show broad correlations with past temperature and rainfall variability, sea-level change and El Niño Southern Oscillation (or ENSO) activity (Williams *et al.* 2008). Those populations collapsed in most regions around 3500 years ago, coinciding with a drier and more variable climate and the onset of a different ENSO regime characterised by more dominant El Niño events. Populations on Australia's arid west coast were also affected (although marine resources provided partial buffering). Subsequently a prolonged period of population expansion occurred over the most recent 3000 years. The rapid decline in population during the last two centuries, of course, had little to do with the climate.

Conclusion

Australia faces the need to clarify and manage domestic demographic-environmental pressures of great portent and increasing urgency. Meanwhile, increasing population pressures in our region along with heightened climatic and other environmental stresses will affect food production. Rice production in Vietnam's prodigious Mekong Delta has already been greatly reduced by sea-level rise and salt-water intrusion. India and China face mounting shortages of freshwater, especially in their northern regions. The Asian Development Bank reported in 2012 that the recent increased frequency of extreme weather events in the Asia–Pacific region had swollen the numbers of people displaced by storms, floods, temperature extremes and rising seas.

We still haven't acknowledged or discussed these connections in currently comfortable Australia. The 2012 government-commissioned report on 'Australia in the Asian Century' focused on economic opportunities, trade, educational exchange and geopolitical security – and largely overlooked the longer shadows being cast by climatic and environmental changes and population pressures. In a Plus 4°C world, now regarded by many scientists as quite likely by 2100, flows of migrants and refugees would increase hugely.

Here we face a dilemma. A fair, humane and cooperatively shared solution to the refugee problem must be found – and Australia, the region's wealthiest large country, should expect to provide much of that sharing of resettlement. That would necessarily involve some further increase in this country's population. So, how to get the balance right between global, regional and national interests – human-centred, not economically centred, interests? To achieve a workable future, we and our elected political 'leaders' should be openly discussing these deeper dimensions of how Australians are managing or mismanaging the issues of population, resources and climatic-environmental changes in the long-term interests of achieving an environmentally sustainable and socially equitable Australia (Raupach *et al.* 2013), including managing the changing complexion of regional stresses and refugee flows. Recent experience of 'angry summers' reminds us that this continent is vulnerable to climatic stresses, currently on track to become more severe.

As in many times over the past 20 000 years and more, and in many parts of the world, a critical juncture between population size, resource base and climatic conditions is now bearing down on Australia's future. As Frank Fenner remarked late in life: 'Our grandchildren will live in a very different and difficult world.'

References

Bennett C, Dear K, McMichael AJ (2013) Shifts in the seasonal distribution of deaths in Australia, 1968–2007. *International Journal of Biometeorology* doi:10.1007/s00484–013-0663-x

deMenocal PB (2001) Cultural responses to climate change during the late Holocene. *Science* **292**, 667–673.

Department of Health and Human Services (Victoria) (2009) *January 2009 Heatwave in Victoria: an Assessment of Health Impacts.* Government Printer, State of Victoria.

Kennett DJ, Breitenbach SFM, Aquino VV, Asmerom Y, Awe J, Baldini JUL, Barlein P, Culleton BJ, Ebert C, Jazwa C, Macri MJ, Marwan N, Polyak V, Prufer KM, Ridley HE, Sodeman H, Winterhalder B, Haug GH (2012) Development and disintegration of Maya political systems in response to climate change. *Science* **338**, 288–291.

McMichael AJ, Woodruff R, Whetton P, Hennessy K, Nicholls N, Hales S, Woodward A, Kjellstrom T (2003) *Human Health and Climate Change in Oceania: A Risk Assessment 2002.* Commonwealth Government, Canberra.

Pinker S (2011) *The Better Angels of Our Nature.* Allen Lane, Penguin Press, London.

Raupach MR, McMichael AJ, Alford KJS, Cork S, Finnigan JJ, Fulton EA, Grigg NJ, Jones RN, Leves F, Manderson L, Walker BH (2013) Living scenarios for Australia as an adaptive system. In *Negotiating Our Future: Living Scenarios for Australia to 2050.* (Eds MR Raupach, AJ McMichael, JF Finnigan, L Manderson, BH Walker) pp. 1–53. Australian Academy of Science Press, Canberra. <http://science.org.au/publications/research-reports-and-policy.html>

Smith MA, Williams AN, Turney CSM, Cupper ML (2008) Human-environment interactions in Australian Drylands: exploratory time-series analysis of archaeological records. *The Holocene* **18**(3), 389–401.

Williams AN, Santoro CM, Smith MA, Latorre C (2008) The impact of ENSO in the Atacama Desert and Australian arid zone: exploratory time-series analysis of archaeological records. *Chungara Revista de Antropologia Chilena* **40**, 245–259.

Williams AN, Ulm S, Cook AR, Langley MC, Collard M (2013) Human refugia in Australia during the Last Glacial Maximum and Terminal Pleistocene: a geospatial analysis of the 25–12 ka Australian archaeological record. *Journal of Archaeological Science* **40**(12), 4612–4625.

14

Climate change – beyond dangerous: emergency action and integrated solutions

Ian Dunlop

Biophysical limits in the 21st century

This chapter is entitled 'Climate change – beyond dangerous'. From a risk management perspective, that is the point we have reached – extremely dangerous territory – as human carbon emissions continue to accelerate on a worst-case path.

The Fenner Conference focused on population and, in that regard, we are also at a unique point. In 1945 we had a relatively empty world of around two billion people. We now have seven billion, in theory heading toward 9–10 billion, with a big question mark as to whether that will actually occur. As a result of exponential increases in both population and consumption, we already have a full world in terms of the impact on the biosphere.

We have moved out of the 20th century where growth has been relatively continuous since World War II. We are now in a period of major discontinuity, negotiating what I term *the rapids of creative destruction*, where the whole concept of growth is in question. We will emerge either with a fundamentally different, sustainable world, or the system as we know it will collapse.

Since World War II, economic growth has been the dominant political mantra. However, in the 21st century we now face totally different, resource-constrained conditions. Unfortunately, the 20th century leaders taking us into this new world are assuming that nothing has changed.

In reality, we face global limits never previously experienced, particularly for energy supply (notably peak oil), climate change, water and food.

These are biophysical limits. They are not determined by economics but by the sheer pressure of consumption on the biophysical system. On average, humanity today needs the biophysical capacity of 1.5 planets to survive, which is clearly not sustainable. If all seven billion lived at Australian levels, four planets would be required, which is even less sustainable!

In turn, this is leading to increasing financial and social instability globally, which is discussed in terms of economic and geopolitical conflict. But the underlying causes are biophysical constraints.

For example, the current conflicts in Egypt and Syria are viewed in terms of economic and religious conflict, but one of the primary causes in both cases is climate change. The Eastern Mediterranean has experienced the most severe drought in recorded history over the last decade, with the result that incumbent totalitarian regimes are no longer able to maintain the subsidisation which has kept their populations relatively docile historically.

These limits are now real and pose major risks to the future of humanity. Here I will confine my comments to climate change and energy, but the risks to food and water security are of equal concern.

Climate change

First, an overview of paleoclimate history, showing global average temperatures going back over the last 65 000 years (Figure 14.1):

Fifty-five million years ago the Paleocene–Eocene Thermal Maximum (PETM) delivered a big spike in temperature. Temperatures then fell, and about 37 million years ago, the Antarctic began icing over. The world cooled further, and about 4.5 million years ago, the Arctic started to ice over.

Over the last 0.5 million years, temperature cycled rapidly before entering the Holocene era of the last 11 000 years. This has been a period of remarkably stable climate, about 15°C plus or minus half a degree C, when humanity as we know it evolved.

Climate deniers argue that this overall picture demonstrates climate has always varied and that what we are seeing today is little more than a continuation of that natural variability. The point they miss is that humanity as we know it did not exist prior to the Holocene.

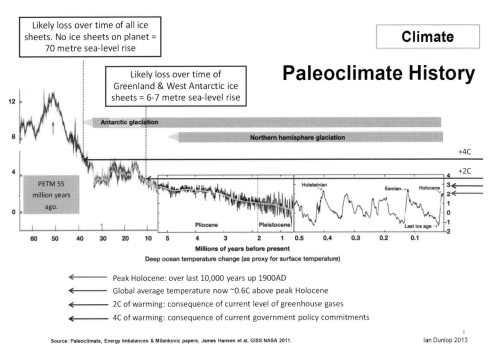

Figure 14.1: Overview of paleoclimate history.

As Figure 14.1 shows, global temperature is now about 0.6°C above the Holocene peak. The official objective is to limit warming to about 2°C above pre-industrial levels. But once equilibrium is reached, that line would take us back to the point where there was no ice in the Arctic. Unfortunately, current global policies fall far short of the 2°C objective and, unless changed, will result in a temperature increase in excess of 4°C. This would mean no ice in either the Arctic or the Antarctic.

The implications are rarely discussed. We assume that 2°C is fine, ignoring the fact that over time it will result in 6–7 m of sea level rise, wiping out cities such as New York, London, Tokyo and Shanghai in their current form. 4°C would lead to around 70 m of sea level rise from the combined effect of the Arctic and Antarctic ice melt, which would be disastrous for humanity given the high proportion of global population living in close proximity to coastlines.

But that is what current policies will deliver, highlighting one of the most difficult aspects of climate change, namely that the inertia of the climate system means we do not see the full effect of our actions today for decades to come. Unless we take far more rapid action now, we are locking in those outcomes; the longer we delay, the harder it becomes to avoid them

So what is the evidence supporting these concerns?

It is clear from this picture of decadal temperature increase over the last four decades that the world is warming, particularly at the poles (Figure 14.2).

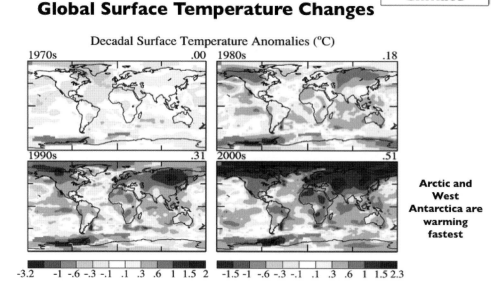

Global Surface Temperature Changes Climate

Decadal Surface Temperature Anomalies (°C)

Arctic and West Antarctica are warming fastest

Decadal mean surface temperature anomalies relative to base period 1951-1980.

Source: update of Hansen et al., GISS analysis of surface temperature change. *J. Geophys. Res.* **104**, 30997-31022, 1999.

Ian Dunlop 2013

Figure 14.2: Decadal temperature increase 1970–2000.

Correspondingly, Arctic sea ice volume, since 1979 has been in rapid decline. The figure shows that decline by month, the lower lines being the warmest months. On current trends, the Arctic would be ice free in summer by around 2015; in winter possibly by around 2030. Since 1979, 80 per cent of the summer sea ice volume has been lost, with half of that loss occurring in the last seven years (Figure 14.3).

Earlier predictions did not anticipate this outcome until the end of the century. So big changes are occurring; it is irresponsible to ignore them, as our leaders are currently doing.

The latest IPCC 5th Assessment Report is an extremely important document. Despite much criticism, it is remarkable what has been achieved. This year's report consolidates previous understandings at a higher level of confidence and highlights risks that have yet to be quantified. For example, sea level rise is conservatively assumed to be about 1 m by the end of the century, but this does not include the potential for more rapid melt of the Greenland ice sheet, as further research is required before a definitive judgement can be made.

One perspective using the limited available data, from NASA's Jim Hansen, is that the melt is accelerating exponentially, which might lead to a 5 m sea-level increase somewhere between 2050 and 2070. Such an outcome is not on the radar at all from a policy-making perspective, yet it is an increasing risk.

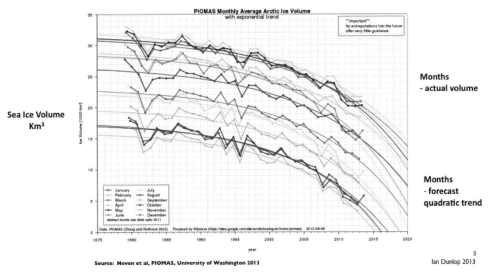

Figure 14.3: Arctic sea ice volume 1975–2020.

Further concerns are potential non-linear climate tipping points, when the climate may flip from one relatively stable state, to another far less conducive to human development. For example, as the white Arctic sea ice melts, reflection of solar radiation decreases and the grey-green sea absorbs more heat. In turn, the permafrost melt speeds up, releasing large amounts of carbon to atmosphere, further accelerating warming.

The Arctic and the permafrost are already showing signs of being close to tipping points. Big changes are also taking place in the Antarctic, which is warming faster than anywhere else on the planet. Numerous other examples of rapid warming are evident around the world.

Catastrophic risk management

Climate change is essentially a matter of risk management, but not in any conventional sense, for example, in the way in which operational or financial market risks are handled. Based on the science and the evidence, which has been before us for some time, it is no exaggeration to say that this is potentially catastrophic risk, in which the entire future of humanity is at stake. We have never faced such risks before; new thinking, and leadership, are required if we are to overcome these challenges.

We probably have passed some climatic tipping points at the 0.8°C warming already experienced, let alone the remaining warming that is now unavoidable due to our historic emissions, which may be around 1.2°C or so. The implication is that we are already locked in to further extreme climatic events, albeit the full effects will not be seen for some time, due to the inertia of the climate system. The evidence clearly indicates that the formal 2°C target is far too high. If we wish to avoid really catastrophic outcomes, we have to initiate an emergency program to reduce emissions far more rapidly than is being discussed officially.

Notwithstanding science and evidence, investors in financial markets continue to regard climate change as a second-order ESG (Environment, Social and Governance) issue, of far less importance than their primary interest of increasing short-term shareholder returns. The markets have yet to grasp that climate change is already a primary driver of the economic system, and of shareholder value. Unless we rapidly find ways to constrain long-term temperature increase some way below 2°C, we will have neither an economy nor a stock market to worry about.

A 4°C world – one billion people

Political and business leaders glibly talk about adapting to a 4°C world with little idea of its implications. It is a world of one billion people or less, not seven billion, caused by a combination of heat stress, escalating extreme weather disasters, sea level rise, disease, food and water scarcity, with consequent social disorder and conflict (Schellnhuber 2009).

As the UK Royal Society put it: 'In such a 4°C world, the limits for human adaptation are likely to be exceeded in many parts of the world, while the limits for adaptation for natural systems would largely be exceeded throughout the world' (Royal Society Transactions UK 2011).

When asked at the Melbourne 4 Degree Conference in July 2011 to explain the difference between a 2°C and a 4°C world, Hans Joachim Schellnhuber, Director of the Potsdam Institute for Climate Impact Research replied simply: 'human civilisation' (Spratt 2011).

Kevin Anderson, Deputy Director of the UK Tyndall Centre for Climate Change Research summarises the dilemma as follows: 'For humanity it's a matter of life or death. We will not make all human beings extinct as a few people with the right sort of resources may put themselves in the right parts of the world and survive. But I think it's extremely unlikely that we wouldn't have mass death at 4°C. If you have got a population of nine billion by 2050 and you hit 4°C, 5°C or 6°C, you might have half a billion people surviving' (Fyall 2009).

'It is fair to say, based on many discussions with climate change colleagues, that there is a widespread view that a 4°C future is incompatible with any reasonable characterisation of an organised, equitable and civilised global community. A 4°C future is also beyond what many people think we can reasonably adapt to. Besides the global society, such a future will also be devastating for many if not the majority of ecosystems' (Anderson 2012).

Large parts of the world would be subject to extreme drought, with severe impact on food and water supply and human health, while other parts would experience intense rainfall and flooding, sometimes both in short order, as per recent Australian experience (McMichael 2011; Dai 2010).

An analogy with human physiology is appropriate. Normal body temperature is 37°C. Add 2°C and you have high fever. Add 4°C and you are probably dead. Just so with the climate.

As the World Bank emphasises: 'If we have any sense of responsibility to current and future generations, a 4°C world is to be avoided at all costs' (World Bank 2012).

Energy

Energy is essential for the development of humanity. The availability of cheap energy from fossil fuels has led to the prosperity we enjoy today, and to much alleviation of poverty around the world. However, fossil fuels are responsible for the bulk of our carbon emissions. Climate considerations now mean that, as energy demand continues to increase exponentially, we have to rapidly wean ourselves off fossil fuels, replacing them with low or zero emission alternatives, as well as slowing, or reversing, that demand growth.

Recent forecasts of primary energy demand, from the International Energy Agency (IEA), demonstrate that without fundamental change to current global policies, by 2035 fossil fuels will still supply close to 80 per cent of requirements. Fatih Birol, Chief Economist of the IEA explained the implications in March 2012: 'The world is currently following a trajectory which will increase temperature by 6°C relative to today, for which the energy sector is largely responsible. If that is allowed to happen, we are all in trouble.'

Figure 14.4 shows fossil fuel consumption from 1850 to the present day. If nothing changes, in the 24 years from 2011 to 2035, we will consume about 70 per cent of all the fossil fuels that have ever been consumed. Given our scientific knowledge of the emission implications, this is simply suicidal.

From another perspective, we reached the peak of conventional, or cheap, oil supply in 2005 and it has plateaued ever since. Production from these oilfields is declining faster than previously anticipated as their reserves are depleted.

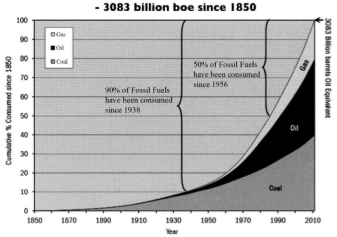

Figure 14.4: Fossil fuel consumption from 1850 to present day.

Accordingly, just to maintain current production, new fields have to be developed, which are typically far more expensive than those they are replacing (Figure 14.5).

In essence, we have to find four new Saudi Arabias by 2035 just to maintain current supply, which is highly unlikely. We are now scraping the bottom of the proverbial world oil barrel, as we turn increasingly to expensive and hard-to-produce unconventional sources of supply.

The peak of global oil supply is the point at which production cannot be increased, irrespective of price or economic factors, because of technical constraints within the oil reservoirs. Production from an individual oilfield roughly follows a bell-shaped curve, increasing initially, plateauing, then declining. If you cumulate all the reservoirs around the world, they demonstrate a similar characteristic. The overall peak in global supply is now only being held off by additions from these unconventional sources.

We are not running out of oil *in toto*. At the peak, we still have available roughly half the oil that will ever be produced from these reservoirs. We also have large unconventional resources. The real issue is how to get those resources to market in a way which is both economically and environmentally sustainable.

Unfortunately the definition of environmentally sustainable has fundamentally changed due to the carbon emission limits determined by climate science. In short,

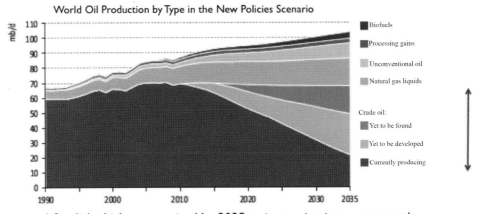

Figure 14.5: World energy outlook 2011.

we can no longer use fossil fuels in the profligate manner to which we have become accustomed. Further, the economics of energy production now have to be assessed in the light of the energy return on energy investment (EROEI), a concept unfamiliar to most economists.

To get energy out, you have to put energy in; in other words drill wells, build pipelines or ships to deliver the product to market. The surplus energy drives society: our cars, our buses, our trains, our cities etc. The graph shows how the ratio of energy-out to energy-in (EROEI) has been declining (Figure 14.6). Historically, in the heyday of cheap oil in the Middle East, it used to be above 50:1. Ever since, the ratio has been dropping and today is below 20:1 globally.

New technologies being developed to access unconventional resources, for example, directional drilling to access shale and extract oil and gas, coal seam gas, are impressive. But this comes at a cost, and their EROEIs are far lower than conventional operations.

To run an industrial civilisation requires a minimum EROEI of about 10:1, and unconventional supply is generally lower than 10:1. Thus, as conventional supply declines, we will no longer have the surplus energy available to run society as we have known it, and conventional economic growth will not be sustainable. The

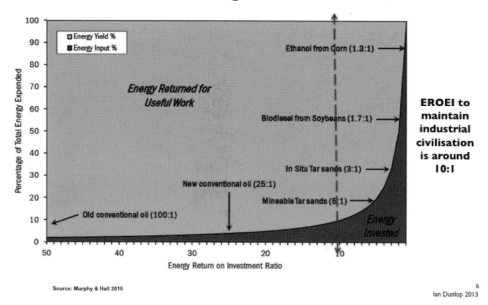

Figure 14.6: EROEI is dropping rapidly.

implications are profound and they are beginning to bite now. Yet such concerns are dismissed out of hand in Australia.

The climate and energy dilemma

Climate and energy are inextricably linked, and integrated solutions are essential if we are serious about sustaining our society while avoiding catastrophic climatic outcomes. They cannot be isolated in separate silos as per current policy-making.

Climate science indicates that if the world is to have a reasonable, say two in three, chance of constraining temperature below the official 2°C increase, albeit 2°C is too high, from now on only about 20 per cent of current proven fossil-fuel reserves can be burnt (Figure 14.7). The remaining 80 per cent must be left in the ground.

This raises further fundamental questions: why then continue to explore for fossil fuels and what value should be placed on resources companies whose future depends on being able to exploit those resources? In risk management terms, our existing proven reserves, if burnt, will generate enough emissions to fry the planet several times over.

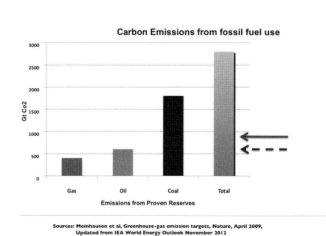

Figure 14.7: Climate and energy are inextricably linked.

The Climate Commission, before it was disbanded, argued that on current trends, the total world carbon budget will be used up by 2028, in 15 years' time.

So what are we doing?

Despite this stark picture, in the 20 years since negotiations on reducing carbon emissions commenced, virtually nothing has been done to curb them, and there are no signs of that occurring via international treaties in the short term. Meanwhile, after a brief pause during the Global Financial Crisis, emissions continue to rise at record rates (Global Carbon Project 2012). These emission growth rates represent a worst-case scenario leading, unless corrected, to temperature increase well in excess of 4°C.

On the other hand, we have official solutions that were supposed to solve this dilemma. Particularly carbon capture and storage – but it is not working. It may make a contribution in due course, but it is not going to do so either to the extent or in the time required, given the risk we now run.

Other so-called clean-coal technologies will not achieve the emission reductions required. The current rush from coal to gas will probably worsen warming. We are also locking in high carbon infrastructure today with a 40–50 year life. Once built, it will require a brave politician to shut it down.

Since World War II, it has taken about 30 years for each major technology change to have a material impact on our energy system. Thus, if we rely on conventional change processes, we will not solve our current dilemma in the time required.

Our refusal to confront these facts has triggered serious discussion in scientific circles on geo-engineering; altering the atmospheric system to prevent warming occurring by, for example, putting sulphates into the atmosphere, thereby replicating emissions from volcanoes.

This is a dangerous development given that the implications are poorly understood. Geo-engineering is a last resort which should not be used while other options are available. However, the inaction of current leaders means that those options are rapidly being cut off.

The policy response

The Australian policy response to this dilemma has been a failure of leadership at all levels.

Most importantly, we are not being honest about the problem, preferring not to 'scare the horses' but talk around the problem without clearly defining it, for reasons of short-term political expediency or self-interest. Problem definition typically provides 80 per cent of the solution, so honesty is essential if you are genuine in your endeavours. It needs to be done carefully, setting out both problem and solution, but it must be done. Otherwise the result is totally inadequate policies that become impossible to change given the political capital invested in creating them, which is exactly what has happened with successive governments.

Business has been ambivalent about acting on the climate science and its risk implications, arguing that it cannot act in the absence of consistent government policy. Simultaneously, vested interests within business have been deliberately disruptive in preventing any serious attempts at climate policy development and implementation.

Major non-government organisations (NGOs) made a strategic mistake some time ago by deciding to work with government without insisting upon an honest articulation of the problem. As a result, they have been locked in to supporting inadequate policies. For this they have been handsomely rewarded with government grants, but it has been highly detrimental to the development of serious climate policy.

Overall, there is a lack of systems-based thinking, in joining the dots of the climate and energy dilemma to form sensible integrated solutions. The result is that the resilience of the Australian economy, and society, to respond to this threat is badly weakened.

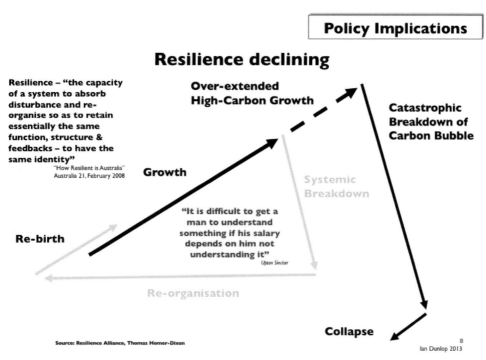

Figure 14.8: System in collapse due to over-extended growth.

Our market economy in the latter half of the 20th century worked remarkably well. Organisations were born, grew, reached limitations, systemically broke down, reorganised and were reborn. That was fine provided there was honesty about the challenges being faced and in developing sensible solutions.

However, once denial of those challenges, and of the real solutions, sets in, the ability of the system to become self-correcting fails, as growth becomes over-extended to the point where the system cannot recover (Figure 14.8). That is now happening with our high-carbon economy as successive leaders in both politics and business, in adopting ideological anti-science agendas, fail to heed the climate and energy warning signs.

This may well end up in catastrophic breakdown – for example, from the bursting of the carbon bubble we have created on the assumption that the 20th century high-carbon world will continue. Business and investors are still rushing into unconventional high-carbon resources, developing projects that are fundamentally unsound on both economic and environmental grounds. The impending collapse will be difficult to recover from as substantial funds will become stranded assets, rather than having been invested in the low-carbon opportunities which represent our future.

The real challenge – an emergency response

Avoiding a 4°C world means emissions in the developed world have to peak within three to four years, then decline rapidly, by about 9 per cent per annum – an unprecedented task that requires nothing less than an emergency response.

It is clear that the existing political, corporate and market economy processes will not deliver – either the level of change we need technologically, or in the time that we require it. So we need a circuit breaker to disrupt conventional thinking.

As Kevin Anderson, Deputy Director of the UK Tyndall Centre for Climate Change Research, put it recently: 'Today, in 2013, we face an unavoidably radical future. We either continue with rising emissions and reap the radical repercussions of severe climate change, or we acknowledge we have a choice and pursue radical emission reductions. No longer is there a non-radical option. Moreover, low-carbon supply technologies cannot deliver the necessary rate of emission reductions – they need to be complemented with rapid, deep and early reductions in energy consumption' (Climate Code Red 2013).

Thus the context of the debate has to change quickly, and we have to be brutally honest about the problem. Whereupon the cry will inevitably go out: 'The government must do something!'

But government is not going to do anything. Experience over the last 20 years tells us that conventional politics is incapable of handling an issue this complex, whether in Australia or globally.

We have to build coalitions of champions, from progressive thinkers who are prepared to speak out from whatever segment of society. For example, community groups, progressive corporates, the military, academia and levels of government who are prepared to face the facts honestly. Also, the supernational organisations, virtually every major supernational organisation – such as the UN, OECD, IEA, World Bank, IMF, and the WEF – are talking in similar terms and they cannot continue to be ignored. These coalitions need to go public, articulating both challenge and solutions, bypassing conventional politics.

Eventually the combination of top-down pressure from progressive thinkers and bottom-up pressure from communities may force governments to respond. It was, after all, a French politician in the 1800s, Alexandre Auguste Ledru-Rollin, who said: 'There go the people I must follow, for I am their leader.'

We must also mandate the critical policy outcomes. It is no longer acceptable to say: 'this is the best we can do, this is all that political realism allows'. We have to define where we are today, where the science tells us we have to get to, and then commit to that path without further procrastination.

It will not be easy, but we know the solutions; we now need the will to implement them. The process must be set outside conventional politics because it will not be successful if it becomes embroiled in that system. This certainly poses issues for democracy, but there are precedents in such emergency situations.

The current situation has major implications for population, as our national carrying capacity is going to be severely constrained by natural events, even if we do take emergency action on climate change. Euphoric plans for population growth will have to be tempered by a realistic assessment of these constraints. At present there is a total disconnect and the dots are not being joined.

Above all, we must have proactive business leadership because corporations, in the end, have to design and implement the solutions. They are well aware of the science; if they're not, they should be.

The role of directors of major corporations is to act honestly, in good faith and, to the best of their ability, in the best interests of their company in perpetuity, not just in the short term – the latter point often gets forgotten. They have a legal responsibility to ensure that the risks their organisations face are identified and that there are systems in place to manage those risks. Businesses claim to be experts in risk management, but this is a risk they have never previously experienced, requiring very different strategy and management techniques.

Conclusion

In a 4°C world, conventional business and politics are not possible. It is in everybody's interests to prevent that happening. We have solutions, but our options are being cut off by inaction and time is extremely short. We need informal groups of progressive leaders to initiate a completely different conversation, building coalitions to trigger genuine emergency action.

By that I mean the sort of initiatives that were taken in the lead-up to World War II, as economies were completely re-orientated in six to 12 months to meet a totally different purpose. That is the challenge that really confronts us.

You already know enough. So do I. It is not knowledge we lack. What is missing is the courage to understand what we know and to draw conclusions. (Lindquist 2007)

References

Anderson K (2012) 'Climate change going beyond dangerous – brutal numbers and tenuous hope'. <http://whatnext.org/resources/Publications/Volume-III/Single-articles/wnv3_andersson_144.pdf>

Climate Code Red (2013) 'Stop tailoring global warming scenarios to make them "politically palatable" says leading climate scientist'. 4 July 2013. <http://www.climatecodered.org/2013/07/stop-tailoring-global-warming-scenarios.html>

Dai A (2010) 'Climate Change: Drought may threaten much of globe within decades'. NCAR: <http://www2.ucar.edu/news/2904/climate-change-drought-may-threaten-much-globe-within-decades>

Fyall J (2009) 'Warming will "wipe out billions"'. *NEWS.scotsman.com*, 29 November 2009.

Global Carbon Project (2012) 'Global Carbon Budget 2012'. <http://www.globalcarbon-project.org/carbonbudget/12/files/CarbonBudget2012.pdf>

Lindquist S (2007) *Exterminate all the Brutes: One Man's Odyssey into the Heart of Darkness and the Origins of European Genocide*. New Press, New York.

McMichael AJ (2011) 'Insights from past millennia into climatic impacts on human health and survival'. PNAS: <http://www.pnas.org/content/early/2012/02/03/1120177109.full.pdf>

Royal Society Transactions UK (2011) 'Four Degrees and Beyond – the potential for a global temperature increase of four degrees and its implications'. <http://rsta.royalsocietypublishing.org/content/369/1934.toc>

Schellnhuber HJ (2009) Scientist: Warming Could Cut Population to 1 Billion. <http://dotearth.blogs.nytimes.com/2009/03/13/scientist-warming-could-cut-population-to-1-billion/>

Spratt D (2011) '4 Degrees Hotter'. Climate Action Centre Primer. <http://www.climateactioncentre.org/sites/default/files/4-degrees-hotter.pdf>

World Bank (2012) 'Turn Down The Heat: Why a 4°C Warmer World Must Be Avoided'. November 2012. <http://climatechange.worldbank.org/sites/default/files/Turn_Down_the_heat_Why_a_4_degree_centrigrade_warmer_world_must_be_avoided.pdf>

Theology confronts global warming and population

Paul Collins

Notwithstanding the utterances of a few church leaders and many fundamentalists denying global warming, most Christian leaders and church members understand the ecological crisis and have embraced the scientific consensus on global warming and environmentalism. In fact the Vatican is the first entirely carbon neutral state on Earth. Sure, it's the smallest state on Earth (the population in 2013 was 839[1] and given that most of its inhabitants are celibate, it is hardly suffering from a population explosion. The Eastern Orthodox Patriarch Bartholomew I (the nominal head of the Orthodox churches) is known as the 'green patriarch', and Popes Francis and Benedict XVI have regularly called for care for the world and made environmentalism central to the social justice teaching of Catholicism. Many Catholic bishops' conferences and Anglican and Protestant church synods have also taken up the challenge.

But there is one thing that stymies most Christians regarding the ecological crisis: they find it almost impossible to confront the issue of population. They are not alone in this. Australia still doesn't have even a suggestion of a population policy and politicians (with the notable exception of Labor's Kelvin Thomson, MHR for Wills in Melbourne) avoid the issue like the plague. David Attenborough says that there 'seems to be some bizarre taboo around the subject' (Attenborough 2011). None of the churches have any developed thought around the issue of population and some more militant social justice activists tend to adopt the shibboleth that anyone discussing this issue is a borderline racist.

Perhaps this is because humankind has never been in this situation before. None of our previous ethical and theological norms apply to the environmental, global warming and population catastrophe facing us. It involves radical rethinking of our oldest and most treasured moral presuppositions. The most important Protestant theologian of the second half of the twentieth century, Jürgen Moltmann, says unequivocally that the ecological crisis 'is not merely a crisis in the natural environment … [but is] nothing less than a crisis in human beings themselves. It is … so comprehensive and irreversible that it can be described … as apocalyptic' (1985, p. xi).

At its most basic we are involved in a shift of cosmologies. By 'cosmology' I simply mean the way in which we understand ourselves and conceive of our relationship with the world around us. Up until the advent of modern science in the mid-19th century all cosmologies were more or less anthropocentric, that is focused on humankind as the essence and meaning of the cosmos. The cosmos was created for us, revolved around us and was interpreted by us.

The cosmology that has underpinned Western culture up until now has essentially been the Judeo-Christian biblical cosmology. In this God views the whole progress of life and history from the perspective of eternity. The cosmos and everything in it begins with creation and lasts until an unknown date in the future which the bible calls the *Parousia*, 'the day of the Lord'. The New Testament sees the primal revelation of God in the person of Jesus Christ, who actually is God; his life is the central event of world history. God's intervention in the world before the Christ-event is recorded in the Hebrew Scriptures (the Old Testament), and the story of the Christ-event and its immediate results are told in the New Testament. This model made sense for most people in Western Christian culture until the mid-19th century when the discoveries of modern science began to open up our contemporary knowledge of the immense age and spatial extent of the cosmos.

Nowadays we know that human history is minuscule in comparison with geological and biological history. While the processes of the cosmos are billions of years old and those of our earth have been developing for about four billion years, *Homo sapiens sapiens* has been on Earth for only 60–100 000 years. We are forced to face our insignificant existence within the whole spectrum of cosmic history. 'We now experience ourselves', Thomas Berry says, 'as the latest arrivals, after some 15 billion years of universe history and after some 4.5 billion years of Earth history. Here we are, born yesterday' (Berry 1988, p. 14). The same applies to spatial expansion: astronomy has shown that we live in an apparently limitless universe. Today the cosmos is viewed as an ongoing energy event, rather than a sudden creation at a specific point in time. It is a dynamic, self-explanatory process that is caught up in its own inner development.

This immediately leads to a crisis in cosmological understanding, which is particularly acute for believers who do not wish to live in a schizophrenic universe. It is especially confronting for Christian theology which has always been particularly anthropocentric. Anthropocentrism is the presupposition that we constitute the entire meaning of the cosmos and that everything was created for us. Thomas Berry says it is rooted in 'our failure to think of ourselves as a species, interconnected with and biologically interdependent on the rest of reality'. We have become besotted with 'the pathos of the human' and take ourselves and our needs as the focus, norm, and arbiter of all that exists.[2] Up until the 16th century the church declared that there were two revelatory manifestations of the Transcendent. The primary one was the natural world (the 'first scripture') and the second was the Bible and church teaching. But from the 16th century anthropocentrism, this second, darker side of biblical religion began to dominate theology. It has reached such a state nowadays that people believe the absurd assumption that the entire world can be turned into a feedlot for humankind with loss of all wildernesses and the extinction of tens of thousands of species.

If anthropocentrism has been a sustaining Christian myth, it assumes a modern form in the secular myth of progress, the notion of infinite development, the idea that limitless human advancement is both possible and desirable. The remote origins of the myth lie in Jewish and Christian notions of an apocalyptic era in which a kind of heaven on Earth can be achieved. In the 19th century this notion became secularised with the theory of evolution playing a major role because, in the popular understanding, evolution was always toward more complex and developed realities. In his 1851 book *Social Statics,* Herbert Spencer, with the lack of self-doubt characteristic of Victorian thinkers, argued that 'progress is not an accident, but a necessity. … As surely as … evil and immorality disappear, so surely must man become perfect' (Spencer 1851).

Nowadays the myth of progress has been transmuted into economic terms with neo-rationalism's concept of a perfect, infinite, ever-growing market, a pseudo-religion if ever there was one! Today one has faith in the Almighty Market and consumption is now a form of divine grace, all underpinned by a cosmology that is anthropocentric and grossly materialistic. Significantly the role of the myth of progress is often forgotten when discussing population, while the failures of the great religious traditions, especially Catholicism, are highlighted.

So what needs to be rethought? First we have to jettison our anthropocentric cosmology. We have to recover the notion that life is an interactive continuum from the most primitive forms to the most highly evolved and complex. We are not separate creatures whose lives and value somehow stand over and against the rest of creation. Berry emphasises that all life is profoundly related genetically. It is the genes that pass on the ever-increasing complexity of life. These genetic relationships constitute a profound oneness. Humankind is not separate and over-against all other reality. It is a constituent part of it. 'Our bonding with the larger dimensions of the universe comes about primarily through our genetic coding. It is the determining factor ... in the organic functioning that takes place in all our sense functions ... in our thought, imagination and emotional life ... [in] our experience [of] joy and sorrow ... It provides the ability to think, speak and create. It establishes the context of our relationship with the divine' (Berry 1988, p. 196). The realisation of our genetic relatedness with everything else will mean that, unless we are prepared to destroy something of ourselves, we will have to work to preserve our common life rather than destroy it. Our relationship to the rest of the cosmos means that species extinction and the destruction of the natural world lessens us, destroys something in us. Our genetic coding defines our humanity and our individuality, as well as constituting our relationship to the rest of reality.

We now live in a limitless cosmos, both spatially and temporally. In this context theology and ethics must embrace the concept that it is absurd to say that humankind constitutes the entire meaning of the cosmos. Our continuing existence is contingent on our recognition that we are a mere blip in cosmic time and space whose existence depends entirely on the rest of creation. We need the humility to recognise we are merely contingent creatures that have been blessed (or cursed?) with an excess of self-awareness and conscience. By attempting to see ourselves in perspective we need to address our driving assumption that we somehow constitute the entire meaning of the cosmos. However, this doesn't mean that our lives are purposeless, but simply that our meaning structure is linked intimately to the whole cosmos.

Interestingly in the Letter to the Romans (8:18–23), Saint Paul has a suggestive apocalyptic theology that is evocative of our biological connectedness to the rest of the cosmos. In this fascinating but difficult text Paul says: 'We know that the whole creation has been groaning in labor pains until now; and not only the creation, but we ourselves, who have the first fruits of the Spirit, groan inwardly while we wait for adoption, the redemption of our bodies' (8:20–23). In this text he links our human fate to that of the whole creation and says that all material reality is struggling through a birth-process that will lead to 'the redemption of our bodies' (8:23). He sees the whole cosmos caught up in a desire-filled movement that will eventually lead to 'the glory about to be revealed' (8:18), with the word 'glory' here essentially referring to God and describes a scene where the whole of nature, in a striking image, is engaged in the process of giving birth to something better. Some commentators argue that here we have Paul laying the foundations for a theology of nature, or at least providing the basis for a more positive interpretation of the world.

Admittedly the passage is complex and has been described by scripture scholars as 'a particularly muddy watering hole' and 'frustratingly allusive' (Hunt *et al.* 2008; Collins 2010, pp. 170–3). While Paul is not a modern ecologist, for him Christ, 'the children of God' and all creation are intimately intertwined. The text unequivocally links the fate of humankind to the liberation of creation. The material world is not merely some type of inert stage or backdrop against which the human drama is played out, but it has an intrinsic purpose and it 'groans inwardly' until it reaches its fulfilment. God cares about creation because God will intervene to liberate and fulfil it. For Paul both creation and humanity are caught up in a process that will lead to 'the redemption of our bodies'. He could just as easily have said 'the redemption of matter' because the Greek word *Sarx* here can mean both 'body' and 'matter', and he envisions a time when the whole material cosmos will reach fulfilment. What the text does tell us clearly is that our fate and that of creation are inextricably bound together as the whole of reality longs for a liberation that only God in Christ can bring.

The moral principle that flows from this is that matter in all its species manifestations has such value that it is destined for transcendent fulfilment. As the medieval theologian Thomas Aquinas succinctly says in the *Summa Theologica*, 'Every creature *demonstrat personam Patris*', which means 'Every creature shows forth the personality of God'. 'Every creature' here means everything in the cosmos from the smallest to the largest and grandest. This principle lays heavy moral responsibilities upon us. If, theologically, the entire cosmos is redolent of the divine, for us to destroy other species and to tear the world apart with our ever-escalating numbers, greediness and belief in an infinite Market involves us in an actual destruction of the image of the Transcendent. This is what might be called a kind of 'deicide', a killing of God. It certainly is a kind of 'biocide'.

Here it is important to remember that theology is not empirical in a scientific sense. Rather it operates at the mythical and poetic level of consciousness and is a kind of metaphor for a reality that is by definition indefinable. So you don't have to agree with my theology to understand that a metaphor or symbol like this can be powerfully used to shift ethical attitudes toward more responsible fertility.

None of this, of course, immediately solves our population problem. Here I am merely emphasising that theology and faith are not *per se* enemies of population policy or control. They could in fact be allies, if only the churches could get beyond their anthropocentrism, and non-believers and secularists their assumption that Christianity, and particularly Catholicism, are benighted and superstitious with nothing to offer this problem. Precisely because it operates at the mythical and unconscious level, theology can change basic human attitudes.

But truth and realism tell me that we have a long, hard haul ahead of us if we are to convince people of this. And global warming time frames are getting tighter and tighter, meaning that as population escalates it becomes ever harder to limit temperature increases and the ever-present danger of tipping points.

Nevertheless I refuse to give up hope. I will quote Saint Paul again, this time in the First Letter to the Corinthians (13:13), in a passage that focuses on love. But I want to change his emphasis slightly. For me the text ought to read: 'And now faith, hope and love abide, these three; and the greatest of these is *hope*.'

Endnotes

1   See: <https://www.cia.gov/library/publications/the-world-factbook/index.html>, accessed 16 June 2014.
2   Personal interview with Thomas Berry broadcast on *Insights*, ABC Radio National, 27 January 1991.

References

Attenborough D (2011) This heaving planet. *New Statesman* **25**(April), 28.

Berry T (1988) *The Dream of the Earth*. Sierra Club Books, San Francisco.

Collins P (2010) *Judgment Day: The Struggle for Life on Earth*. UNSW Books, Sydney.

Hunt C, Horrell DG, Southgate C (2008) An environmental mantra? Ecological interest in Romans 8:19–23 and a modest proposal for its narrative interpretation. *Journal of Theological Studies* **59**, 546–579.

Moltmann J (1985) *God in Creation: An Ecological Doctrine of Creation*. SCM Press, London.

Spencer H (1851) *Social Statics; on the Conditions essential to Happiness specified, and the First of them Developed*. John Chapman, London. (From the Online Library of Liberty at <libertyfund.org> chapter 2, para 4)

16

Denial as a key obstacle to solving the environmental crisis

Haydn Washington

Denial is not just a river in Egypt. In fact it is arguably the greatest problem in the human psyche. Why? Because it makes us a seriously dumb species. It turns off our intelligence so we don't use our creativity to solve major problems. There are (at least!) *four huge elephants* in the room that society mostly doesn't see: population, consumption, the growth economy and climate change. All of these have been ignored and denied by the majority of governments (and 'we the people') over many decades. As a society, we continue to act as if there is no environmental crisis, no matter what the science says. How is it possible for civilisations to be blind toward grave approaching threats to their security? Denial is as old as humanity, and possibly nobody is free from it (Zerubavel 2006). We proceed often in a cultural trance of denial, where people and societies block awareness of issues too painful to comprehend. This human incapacity to hear bad news makes it hard to solve the environmental crisis. We call ourselves *Homo sapiens*, but many of us seemingly are actually *Homo denialensis*.

Scepticism vs. denial

But what *is* denial, is it the same as scepticism? The *Oxford English Dictionary* definition of a sceptic is:

> *A seeker after truth; an inquirer who has not yet arrived at definite conclusions.*

So we should *all* be sceptics in many ways, as we should all seek the truth. Genuine scepticism in science is one of the ways that science *progresses*, examining assumptions and conclusions. 'Denial' is something different, a refusal to believe something no matter what the evidence. Denial isn't about searching for truth, it's about the denial of a truth one doesn't like. So scepticism and denial are in many ways *opposites*. Scepticism is healthy in both science and society – denial is not.

Denial is common

We deny some things as they force us to *confront change*, others because they are just too painful, or make us afraid. Sometimes we can't see a solution, so problems appear unsolvable. Thus many of us deny the root cause of the problem. Psychoanalysis sees denial as an 'unconscious defence mechanism for coping with guilt, anxiety or other disturbing emotions aroused by reality' (Cohen 2001). Zerubavel (2006) notes that the most public form of denial is 'silence', where some things are not spoken of. He notes that 'silence like a cancer grows over time', so that a society can collectively ignore 'its leader's incompetence, glaring atrocities and impending environmental disasters'. He concludes that denial is inherently *delusional* and inevitably distorts one's sense of reality.

People often get upset when confronted with information challenging their self-delusion. Many prefer illusions to painful realities, and thus cherish their 'right to be an ostrich'. However, Zerubavel (2006) notes that the longer we ignore 'elephants', the larger they loom in our minds, as each avoidance triggers an even greater spiral of denial. The environmental crisis has now got to the point where the elephant is all but filling the room. We may now at times 'talk about' it, but we still deny it. Denial, however, can become a 'pathology' when it endangers the ecosystems humans rely on (Washington 2013).

The history of denial

Denial is as old as humanity. Examples of historical denial are wilderness, population, DDT, nuclear winter, tobacco, acid rain, the hole in the ozone layer, the biodiversity crisis and climate change (Washington 2013). The climate change 'denial industry' has been detailed in Monbiot's (2006) *Heat*, Hoggan's (2009) *Climate Cover Up*, Oreskes and Conway's (2010) *The Merchants of Doubt*, Washington and Cook's (2011) *Climate Change Denial: Heads in the Sand* and Norgaard's (2001) *Living in Denial*. There is far *more* involved than merely 'confusion' about the science. There is a deliberate attempt to confuse the public, so that action is delayed. The denial industry deliberately seeks to sow doubt about environmental problems (Oreskes and Conway 2010) and solutions such as renewable energy. Denial books on the environmental crisis continue to emerge. Jacques *et al.* (2008) explain that between 1972 and 2005 there were some 141 denial books published, of which 130 came from conservative 'think tanks'. In the 1990s, 56 denial books came out, 92 per cent linked to right wing foundations or think tanks (Oreskes and Conway 2010).

Ideological basis for denial

Oreskes and Conway (2010) detail the support conservative think tanks give to denial. The link that united the tobacco industry, conservative think tanks and a group of denial scientists is that they were *implacably opposed to regulation*. They saw regulation as the slippery slope to socialism. They felt that concern about environmental problems was questioning laissez-faire economics and free market fundamentalism. These conservative bodies equate the free market with 'liberty', so if you attack the market then you attack liberty, and hence must be denied – and the science along with you (Oreskes and Conway 2010). The basis for much denial is thus not science but ideology.

Psychological types of denial

There is more involved than just the denial industry. Cohen (2001) lists three types of denial:

Literal denial – the assertion that something did not happen or is not true. For climate change, this is fossil fuel companies saying warming is not caused by humans, or has ceased.

Interpretive denial – the facts are not denied, but are given a different interpretation. Euphemisms and jargon are used to dispute the meanings of events (e.g. 'collateral damage' and not killing citizens). It can also be called 'spin', and is common in governments and business.

Implicatory denial – where what is denied are 'the psychological, political or moral implications … Unlike literal or interpretive denial, knowledge itself is not at issue, but doing the "right" thing with the knowledge' (Cohen 2001). People have access to information, accept this information as true, yet for a variety of reasons, *choose to ignore it* (Norgaard 2011).

Implicatory denial is common in the public. Much of the knowledge about an environmental problem is accepted, but fails to be converted into action (Cohen 2001). 'Distraction' is also an everyday form of denial. If we are worried about something, we tend to 'switch off' and shift our attention to something else. We can also 'de-problematise' it by rationalising that 'humanity has solved these sort of problems before' (Hamilton 2010). We can also 'distance ourselves' from the problem by rationalising 'it's a long way off'. There is also 'hairy-chested denial', where people deny climate change, since it will impact on pleasures such as fast cars. 'Blame-shifting' is another part of implicatory denial, where we shift the blame onto others, such as the US, China or industry (Hamilton 2010).

Ignoring the elephants

Society is very good at not seeing these four elephants. Meyerson (see Hartmann *et al.* 2008) notes re population:

> *Conservatives are often against sex education, contraception and abortion and they like growth – both in population and in the economy. Liberals usually support individual human rights above all else and fear the coercion label and therefore avoid discussion of population growth and stabilisation. The combination is a tragic stalemate that leads to more population growth.*

Challenging consumerism is often seen as *challenging the growth economy* – a key given truth of modern society. Consumption for some has become the meaning of life (constantly promoted by advertising). People are finally starting to *see* climate change, and hence cannot completely ignore it. However, possibly the hardest elephant of all for society to 'see' is the growth economy (Dietz and O'Neill 2013). Unless we can overcome denial and see *all* elephants, we will not solve the environmental crisis.

Types of denial arguments

Diethelm and McKee (2009) explain there are five types of denial arguments:

1   cherry picking
2   fake experts
3   impossible expectations
4   misrepresentations and logical fallacies
5   conspiracy theories.

Denial arguments keep coming. For example, in 2009 there were some 80 climate change denial arguments, now the website <www.skepticalscience.com>

lists 174. So assess denial arguments, and understand their nature (so you can better debunk them).

Breaking the denial dam

Rees (2008) concludes that on the dark side of myth, our shared illusions converge on deep denial. Our best science may tell us that the consumer society is on a self-destructive path, but we successfully deflect the evidence by repeating in unison the mantra of perpetual growth. The first step toward a more sustainable world is to accept ecological reality and the socio-economic challenges it implies. While there is a tendency in society to deny things, there is also a tendency to challenge denial (Zerubavel 2006), and it is this we must foster. We *can* break the denial dam.

And solutions exist (e.g. Washington 2013). Engelman (2012) lists nine humane ways to stabilise population. Population Media (<www.populationmedia.org>) has success promoting population stabilisation. Consumerism was deliberately constructed (or at least massively expanded) after World War II. It needs now to be *deliberately deconstructed* (Assadourian 2013). We can ban 'planned obsolescence' and make 'cradle to cradle' products *mandatory*. We can introduce an advertising tax to rein in the $500 billion yearly spent on promoting over-consumption (Assadourian 2013). Consumption that undermines wellbeing should be discouraged (e.g. by 'choice editing', Assadourian 2013) and we should replace private consumption of goods with public consumption of services (e.g. libraries, public transport). We can adopt a low carbon and low material use 'green' economy (UNEP 2011) immediately. Then we should move to a *steady state* economy, where population and throughput of energy and materials are stable and sustainable (Daly 1991).

So *how to rebut denial*? First, focus on those genuinely confused. It is almost impossible to change those in strong denial. Lord Molson stated: 'I will look at any additional evidence to confirm the opinion to which I have already come' (Tavris and Aronson 2007). Second, lead with positive facts and supply a narrative of how the denial argument misleads. In regard to climate change, explain that every Academy of Science and 97.5 per cent of practising climate scientists (Cook *et al.* 2013) are *saying the same thing*. Explain we need to apply the 'Precautionary Principle' to protect future generations. Australia is one country at major risk from climate change. This chapter has covered a lot of ground quickly, and those interested in digging deeper should see *Climate Change Denial: Heads in the Sand* (Washington and Cook 2011).

To conclude, denial has probably been with us since we first evolved. We let ourselves be duped, we let our consciences be massaged, and we let our desire for the safe and easy life blot out unpleasant realities. We delude ourselves. It is time to wake up, and *break the denial dam*. If a large part of the public abandons denial, they can fairly quickly turn around corporate denial (especially if it costs businesses profits). If people tell our politicians that they want real action (not weasel words) then politicians will actually *act* (Washington 2013). We are not powerless drones who cannot change things. En masse, if we accept the task of repairing the Earth, we have the vision, the creativity and the power to solve the environmental crisis, and to move to a truly sustainable future. That nobody should deny.

References

Assadourian E (2013) Re-engineering cultures to create a sustainable civilization. In *State of the World 2013: Is Sustainability Still Possible?* (Ed. L Starke). Island Press, Washington, DC.

Cohen S (2001) *States of Denial: Knowing About Atrocities and Suffering*. Polity Press, Cambridge.

Cook J, Nuccitelli D, Green S, Richardson M, Winkler B, Painting R, Way R, Jacobs P, Skuce A (2013) Quantifying the consensus on anthropogenic global warming in the scientific literature. *Environmental Research Letters* **8(2)**, 1–7.

Daly H (1991) *Steady-State Economics: Second Edition with New Essays*. Island Press, Washington, DC.

Diethelm P, McKee M (2009) Denialism: what is it and how should scientists respond? *European Journal of Public Health* **19**(1), 2–4.

Dietz R, O'Neill D (2013) *Enough is Enough: Building a Sustainable Economy is a World of Finite Resources*. Berrett-Koehler Publishers, San Francisco.

Engelman R (2012) Nine population strategies to stop short of 9 billion. In *State of the World 2012: Moving Toward Sustainable Prosperity*. (Ed. L Starke). Island Press, Washington, DC.

Hamilton C (2010) *Requiem for a Species: Why We Resist the Truth about Climate Change*. Allen & Unwin, Australia.

Hartmann B, Meyerson F, Guillebaud J, Chamie J, Desvaux M (2008) 'Population and climate change'. *Bulletin of Atomic Scientists*, 16 April 2008, see: <http://www.thebulletin.org/web-edition/roundtables/population-and-climate-change>

Hoggan J (2009) *Climate Cover Up: the Crusade to Deny Global Warming*. Greystone Books, Vancouver.

Jacques P, Dunlap R, Freeman M (2008) The organisation of denial: conservative think tanks and environmental scepticism *Environmental Politics* **17**(3), 349–385.

Monbiot G (2006) *Heat: How to Stop the Planet Burning*. Penguin Books, London.

Norgaard K (2011) *Living in Denial: Climate Change, Emotions, and Everyday Life*. MIT Press, Massachusetts.

Oreskes N, Conway M (2010) *Merchants of Doubt: How a Handful of Scientists Obscured the Truth on Issues from Tobacco Smoke to Global Warming.* Bloomsbury Press, New York.

Rees W (2008) 'Toward sustainability with justice: are human nature and history on side?' In *Sustaining Life on Earth: Environmental and Human Health through Global Governance.* (Ed. C Soskolne). Lexington Books, New York.

Tavris C, Aronson E (2007) *Mistakes Were Made (But Not By Me): Why We Justify Foolish Beliefs, Bad Decisions, and Hurtful Acts.* Harcourt Books, Orlando, Florida.

UNEP (2011) *Towards a Green Economy: Pathways to Sustainable Development and Poverty Eradication.* United Nations Environment Programme, see: <www.unep.org/greeneconomy>

Washington H (2013) *Human Dependence on Nature: How to Help Solve the Environmental Crisis.* Earthscan, London.

Washington H, Cook J (2011) *Climate Change Denial: Heads in the Sand.* Earthscan, London.

Zerubavel E (2006) *The Elephant in the Room: Silence and Denial in Everyday Life.* Oxford University Press, London.

17

Why can't we win on population?

Kelvin Thomson

One Saturday night recently I went to my School Reunion for the class of 1972 and 1973. Forty years on. I have spent 25 of them as a state and federal MP – some of my schoolmates thought this a senseless waste of human life! But I was more interested in their lives. One had dropped out of University after a couple of terms and within 48 hours walked into a job as a trainee medical scientist, around which he has built a lifetime career. One girl never completed Year 12, and when she dropped out of school, her mum took her to a local hospital and straightaway got her a job as a nurse. She is still nursing. Others went into business and are now semi-retired or fully retired. One said to me: 'I was so lucky to be born when I was. If I'd come along today, I'd just be a bum.' Well, I don't know about that bit, but he was certainly right about being born in the 1950s. Because the opportunities that my generation had – job and career opportunities, housing opportunity, free education – our children don't have them. For all the hype about growth and progress and development building a better world, it isn't. It is way tougher for our kids than it ever was for us.

This is not just true for Australia; it is true in many other countries as well. It is heartbreaking to hear those stories of all the African migrants who drowned off the coast of the Mediterranean island of Lampedusa. Terrible, terrible, terrible. There is a response that says we should tackle this problem by dismantling our borders and allowing people to live wherever they want to live. But anyone who has seen the 'Gumballs' video by Roy Beck, of Numbers USA – and if you haven't, I can't recommend it too highly – will know that there are two billion people in the world living on $2 per day or less, and that their numbers are increasing by 80 million every year (Beck 2010). No nation in the world – not the United States, not Europe, not Australia – can cope with such numbers.

There are two causes of mass migration. One is people fleeing political violence and repression. The other driver is poverty and people wanting a better life.

In those countries that are beset by political violence, the most common cause is religious fundamentalism. There is religious violence, oppression of minorities, not enough respect for the rights of women, and not enough separation between religion and politics, between Church and State. This needs to be called out. It is a task for all of us – from whatever religious or ethnic background we come – to condemn, to denounce, to shun, to treat as outcasts religious leaders who preach hate and violence. It has to be called for what it is. Until political and religious violence stops, there will be people fleeing it.

And in the other motive for getting on board a boat – the search for a better life – again we all have a role to play. We heard yesterday that each new arrival in Australia creates an infrastructure cost of $340 000, which would go a long way towards lifting an entire African village out of poverty (see Lindenmayer, Chapter 2 this volume).

We should lift our foreign aid budget to 0.7 per cent of GDP. We should not cut our aid by $4.5 billion over the forward estimates as the Liberal Government is doing. It is claimed there is a budget emergency and we can't afford this aid. Then why is the defence budget to be increased? The Government target of 2 per cent of GDP is quite arbitrary and absolute nonsense. Spending money on aid builds goodwill with our neighbours and makes us more secure – I've seen it with my own eyes – people in Indonesian villages like us. In stark contrast spending money on more powerful weapons just makes our neighbours suspicious and sets in place a vicious circle of arms race, fear and mistrust.

The question I have heard at this conference is: why don't we win? Why is Australia's population not only still increasing, but increasing by more than it used to? Why has Australia's net migration rocketed up from 80 000 in the mid-1990s to over 200 000 nowadays? After all, all the opinion polls regularly show that between two-thirds and 70 per cent of Australians don't support the high population, high migration path we are on.

It is worthy of note that no Australian political leader has ever gone to an election promising to increase the migration intake. The Whitlam government reduced immigration to just over 50 000. In more recent times migration increased substantially, during the final Howard years, and again under Kevin Rudd, who declared himself a fan of a Big Australia and achieved no electoral success after that. Julia Gillard declared herself an opponent of Big Australia but the migration rate and population growth continued largely unchanged.

So why don't we win? A central reason is that there is virtually no public debate about it. It is given no oxygen – snuffed out. Why is it so? After considerable reflection, I have come to the conclusion that population is not unique in this regard. It is one of a number of issues, not the only one, but one of a number of issues, that are considered threatening to the economic interests of the wealthiest and most powerful Australians (and in some cases non-Australians) who exercise great influence on our political debate through their direct and indirect media influence.

There are political issues that contain no germ of threat to corporate wealth – same sex marriage, asylum seekers, the Republic, politicians' entitlements. These occupy endless column inches and airtime. If they distract and divide us, so much the better.

But issues that have the potential to impact on the wealth of the wealthy – executive salaries, trade practices and market concentration, foreign ownership, threats to the environment from industry and agriculture, and yes, population growth and migration – these issues are constantly overlooked and repressed.

So one of the key reasons we lose is that we cannot get going or sustain a public debate about this issue. In this we don't get any help from quite a few people who think of themselves as progressive, and who would look you in the eye and swear black and blue that they want to save the environment, they want to protect workers, that they care about the future. But whether it is from fear of being called racist or xenophobic, or a form of moral conceit or vanity, they will not touch the issue of population.

That is, of course, their right, but let me make this point to such people as bluntly as I can. For as long as Australia's rapid population growth, high migration path endures, it will destroy the things you claim to hold dear. It creates a surplus pool of labour, which is used as a battering ram against job security, and against workers' pay and conditions. It prevents us attaining full employment, and the quest for jobs, jobs, jobs for our increasing workforce leads us to sacrifice our environmental standards, destroy wildlife habitat and compromise our quality and way of life. It undermines what you say you are trying to achieve, and assures victory to your political opponents.

And I think honourable defeat is overrated. In my view there is little honour in avoidable failure. In my view those who want to save the environment, who want to help workers, who care about the future, have a responsibility to succeed. Honour comes from success, from solving problems, from being able to proudly hand the baton over to the next generation.

So what am I doing about it? I did not contest the Shadow Ministry election in October. I was a Shadow Minister for 10 years during the Howard years; I have been there and done that. But I am certainly not looking for the quiet life and to slip quietly out the back door.

It is precisely in order to focus on the things that really matter that I have taken the steps I have. I will devote myself in this Parliament to doing everything I can to get the neglected issues, like population, considered.

I admit that to date I have not been all that successful with my efforts. Elements of my 2009 '14 Point Plan' have been adopted – the Baby Bonus is gone, the Labor Government lifted the refugee intake to 20 000 – so I will at some point re-write the Plan (Thomson 2009). But I look at our population growth statistics and projections, and I look at our vanishing birds and plants and animals, and feel that I have been essentially unsuccessful.

So I have set up an NGO – an Incorporated Association – to pursue the cause. It has a token membership fee and is not a competitor organisation for Sustainable Population Australia or the environment NGOs. It is not a political party; I am in one of those already! It is not a single-issue group – I don't think politicians feel under much pressure from those.

It does not try to have a policy on everything, and it tries to avoid the divide and distract traps I talked about earlier. But it sets out a coherent, superior, alternative to the path we are on.

I launched it on 1 December. It is called Victoria First and limited to Victoria, which will disappoint some of you, but you've got to start somewhere. I may be able to work up Associate Membership for interstaters. The great 19th-century philosopher John Stuart Mill said:

> *Solitude, in the sense of being often alone, is essential to any depth of meditation or of character, and solitude in the presence of natural beauty and grandeur, is the cradle of thoughts and aspirations which are not only good for the individual, but which society could ill do without. Nor is there much satisfaction in contemplating the world with nothing left to the spontaneous activity of nature. ... every hedgerow or superfluous tree rooted out, and scarcely a place left where a wild shrub or flower could grow without being eradicated as a weed in the name of improved agriculture. (Mill 1848)*

People who are interested in joining Victoria First can contact Julianne Bell at <jbell5@bigpond.com>. I hope you will help me build it into a large movement of citizens dedicated to passing on to our children and our grandchildren a world in as good a condition as the one our parents and grandparents gave to us.

References

Beck R (2010) Immigration, World Poverty and Gumballs. <http://www.youtube.com/watch?v=LPjzfGChGlE>

Mill JS (1848) Of the stationary state, from *Principles of Political Economy* (extracts online) <http://www.panarchy.org/mill/stationary.1848.html>

Thomson K (2009) There is an alternative to runaway population – Kelvin Thomson's 14 point plan for population reform. <http://www.kelvinthomson.com.au/Editor/assets/pop_debate/091111%20population%20reform%20paper.pdf>

18

Thinking at species level

Julian Cribb

The solutions to human population growth and the challenges of sustainability and climate change may well lie in the most import event in human evolution in the past two million years.

I refer, of course, to the linking of human minds, values, information and opinion at lightspeed and in real time around the planet, via the internet and social media.

This is a development without precedent, not only in our own history but also among all the species which have inhabited or currently inhabit the Earth.

Almost unawares, it is giving birth to an entirely different kind of humanity.

In the second trimester of a baby's gestation a marvellous thing happens. The neurons, axons and glia in the embryonic brain begin to connect – and cognition is born.

An inanimate mass of cells becomes a sentient being, capable of thought, imagination, memory, logic, feelings and dreams.

Today individual humans are connecting, at the speed of light, around a planet – just like the cells in the foetal brain.

We are in the process of forming, if you like, a universal, Earth-sized mind. What Teilhard de Chardin called the noosphere is becoming incarnate (de Chardin 1955, p. 191ff).

A higher understanding, and potentially a higher intellect, is in genesis – capable of interpreting, and potentially solving, our problems at *supra-human level* by applying millions of minds simultaneously to the issues, by sharing knowledge freely and by generating faster global consensus on what needs to be done.

At the very moment in our social evolution when our governments and existing institutions are failing to tackle the overwhelming issues of population, resource scarcity, pandemic poisoning and climate change, a new form of human interconnection and self-awareness has emerged that, just possibly, might save us from ourselves.

A million years ago, as we sat around the campfire on the African savannah, humans ensured our own survival by forging a complex society capable of identifying and overcoming the many threats that then surrounded us.

That experience has carried through our entire existence, and is the mainspring of the creature we have become. We are very, very good at spotting risks and dangers, and finding collective ways to mitigate them. You could say that threat avoidance is our greatest single attribute.

As a long-time newspaper editor I know that the media likes bad news, not because it is naturally vicious, but *because the public demands it*. If you print only good news in your paper, your readership goes down. I have done the experiment, with a real newspaper.

So whence this public demand for news about crises, cancer scares, disasters, horrors, crimes and other unpleasantness? According to the 'Cribb hypothesis' it is our million-year-old habit of looking out for potential risks and threats – and then figuring out how to avoid them – that is at work.

Today we are engulfed with risks, the result of our burgeoning population and the overgrowth in its demands on the Earth's natural resources and systems. Only a fool would think we can keep on doing this without grave existential risk to our entire civilisation and maybe our species.

However, solutions to all of these problems exist.

The solution to population growth is already being implemented by the young women of the world, who are declining to marry and have babies in *all* societies. They are ignoring what men tell them. They are ignoring the priests, the journalists, the politicians, the government baby bribes. They are acting autonomously to reduce their own fertility.

Something very great and very beautiful is happening among women, at the species level. Given worldwide support and approbation, women can bring the population back to around four billion by the early 22nd century *voluntarily*.

The solution to resource scarcity is recycling. As the population falls, there will be no more need of mining – all the metals, nutrients and materials we will ever need will be available and readily accessible in the waste stream. We just need to mine that and recycle them endlessly.

The solution to our money-driven system is to dematerialise wealth – to build an economy founded on products of the mind, rather than material goods. To employ people in creative industries, rather than anachronistic work like manufacturing, mining and agriculture (which will be performed robotically). This way, money will not be used to destroy things of material value such as soil, water, biodiversity and the atmosphere, as it is presently being allowed to do. Money, being immaterial and infinite, can be used to create products that are equally immaterial and infinite – products of the human mind and imagination, which are where the true future wealth of society will reside.

The solution to both climate change and to the pandemic poisoning of all humans and life on Earth is also plain – we can achieve both by eliminating our use of oil, gas and coal, and embracing renewable energy and algae technology for fuel, food, fibre, industrial chemicals, plastics and medications.

The problem we face is that no government in the world is likely to support such a program wholeheartedly. They are dinosaurs, bogged in a tar-pit of empty opinion polls and myopic vested interests which resist change.

So how do we solve these mega challenges?

By 2017 there will be 3.6 billion internet users on Earth and by the 2040s most of humanity will be online (Hettick 2013).

Young people are reaching out to one another in real time, across the divides of race, nationality, ethnicity, religious belief and prejudice. They are learning how alike we all are. How many things we share. How we can 'like', help, support and depend on each other.

They are also learning how deadly are the prejudices, the ignorance, the fears and the hatreds of their parents towards other humans. And how utterly pointless.

The antidotes to ignorance and fear are knowledge and understanding. The internet can supply both.

Humanity is still in the second trimester of the formation of a universal mind – a connected humanity capable of collective thought, information sharing and resolute collaborative action.

It is far too early to dismiss this development, as some might be tempted to do if they are not fully aware of the power of social media and the web. Be sceptical, by all means, but also be open to the possibilities.

Dream the dream of humanity starting to think together at the species level.

As a species we are still in the pre-school of learning to think together, of developing the universal reasoning capable of understanding and solving our common challenges.

Social media, which many people still regard as trivial and of no account, are taking over from traditional forms of controlled media and politics in extraordinary ways. They are a grand reflection of our common values, as well as our vices. They are used by Barack Obama and the United Nations as well as rock stars and airheads.

Out of this inchoate planetary chatter, common threads of thought are already emerging.

Through the internet, knowledge once held only by elites is being exchanged, values shared, attitudes reshaped – and from this, gradually, a worldwide consensus for action will form.

If such a consensus of the majority of the Earth's citizens starts to emerge on *any* of the major existential threats that confront us, it will be an act like no other in history.

It will be one that no government, no corporation, institution or society can ignore.

It will be more powerful than nations or governments – because they will have no power over it.

Its influence will exceed that of the great religions.

It will be more potent than huge corporations. As we are already seeing in issues such as slavery and ethical consumerism, the views and values of millions of concerned consumers will change the way industry behaves, the products and services it produces and the rules by which it operates.

This will apply market forces in unprecedented ways to clean up our energy system, our food supply and our toxically chemicalised planet.

It will do this by exerting a force that many scientists profess to despise: fashion.

Fashion is not just about smart clothes and trendy adornments. It is also about ideas and values expressed in our consumer choices and the lives we choose to lead, the technologies we prefer, our political opinions. It attracts mass attention because it is the bow-wave of change and innovation. It can be about serious and important changes, as well as trivial and flippant ones.

If having fewer babies, seeking cleaner food and healthier products, and rejecting fossil fuels become universal fashions among the young, and are followed by billions of people interested in their own future, it will change how global society is regulated and the signals it sends to both industry and government.

Such a consensus will be more influential on the human destiny than any power or principality heretofore.

Can we make it work? The simple answer is that, if we don't, then fewer than a billion humans will inhabit the famine-, disease-, climate- and war-ravaged ruins of our planet 100 years from now (Schnellnhuber 2009). We have the strongest of all possible motives to succeed. Survival.

This is the time when we get to choose whether we are truly *Homo sapiens* – or some other organism, that failed the Darwinian test.

References

de Chardin PT (1955) *The Phenomenon of Man*. Harper Perennial, New York.

Hettick L (2013) Cisco study projects 3.6 billion Internet users by 2017. *Network World Fusion* **31**(May), 3.

Schnellnhuber HJ (2009) Scientist: Warming Could Cut Population to 1 Billion. See: <http://dotearth.blogs.nytimes.com/2009/03/13/scientist-warming-could-cut-population-to-1-billion/?_r=1>

19

Nor lose the name of Action!

Roger Short

How Frank Fenner would have loved this Sustainable Population Australia conference on Population, Resources and Climate Change! Frank was far ahead of his time, since way back in 1969 he was convinced that the biggest problem the world faced was too many people. Realising that we might even bring about our own extinction as a result of our activities, he bemoaned the fact that our politicians are woefully ignorant of the findings of the Life Sciences (Boyden *et al.* 2013). Amen to all that.

Since 1969, the situation has become far worse, as evidenced by all the speakers at this conference. When I was born in 1930, there were only two billion people on Earth; today, there are over seven billion, and we are growing by one million every five days. The latest projections suggest that we will have reached 10 billion by the end of this century (Emmott 2013). On a per capita basis, little Australia is probably the most affluent and the most effluent nation on Earth – a mixed blessing if ever there was one! But we are wise enough to make a difference that could change the world. On 27 December 2013, the Australian Bureau of Statistics estimated that Australia's resident population was 23 330 682, with an annual growth rate of 1.8 per cent, 40 per cent due to natural increase and 60 per cent to immigration (ABS 2013). Is it not time that one of the world's fastest growing, wealthiest and most polluting developed countries took some action?

The most recent detailed review of the global population is to be found in the Population supplement published in *Science* in 2011, where Australia gets barely a mention (Chin *et al.* 2011). Surely the whole world has much to learn from China's 'One child family' policy launched by Chairman Mao in the early 1980s. It has transformed China from being the most densely populated country in the world with a poverty-stricken rural economy, into an economic urban powerhouse with a relatively stable population of around 1.3 billion – something that could only have been achieved by an autocratic, authoritarian government. Compare this to the situation in poverty-stricken India, whose population will overtake that of China by about 2020. For how much longer can it afford a democratic government?

One or two facts stand out. The highest rates of population growth are in the poorest developing countries of the world. In sub-Saharan Africa, desired family size is high, and conflict and famine are everywhere to be seen. But Africa is the motherland of mankind; how can we bear to see our kith and kin struggling to live on under $2 a day? Were it not for our selfish genes, we would never accept such a situation. Even in developed countries, with small families, there is growing concern that there may not be enough young people in employment to support the ever-increasing number of the elderly in their retirement. But who says we should retire at 65? I am still working full-time at the age of 83, although I no longer get paid for the privilege.

So let us imagine a new future for our species, one that would enable us to live at peace with one another with a stable population and an adequate food supply, without polluting the environment. How could that be achieved?

One of my mentors was the late Sir Dugald Baird, the Professor of Obstetrics in the University of Aberdeen, who in 1965 published a memorable paper entitled 'A Fifth Freedom?' (Baird 1965). In addition to the four freedoms enunciated by President Roosevelt in 1941 – freedom of speech and expression, freedom to worship God in your own way, freedom from want, and freedom from fear – Sir Dugald suggested a fifth one: freedom from the tyranny of unwanted fertility. What a brilliant idea! If we could give all the women of the world free access to contraception, abortion and sterilisation to regulate their fertility as they wished, their wisdom would ensure that 'A Plague of People' was a thing of the past. But it would require a small, affluent, intelligent nation to pioneer such a change. It would be vigorously opposed by the pharmaceutical industry who would resent the loss of profit, the medical profession who would lose influence, and male-dominated religions with their sexist moralities. But Australia, you could do it!

Where would we start? Take the oral contraceptive pill off prescription and have it on sale over-the-counter in all pharmacies. We could even manufacture it in Australia, since it is off patent. Instead of having one of the highest teenage pregnancy and abortion rates in the developed world, we could follow the Dutch example, and make contraception available to schoolchildren before they become sexually active. We would need to re-educate the teachers, but we have already shown in our Teach The Teacher program that final year Medical students are an excellent resource for teaching Education students, and it costs nothing (<www.teachtheteacher.com.au>).

So, what are we waiting for? Action!

References

Australian Bureau of Statistics (2013) Population clock. <http://www.abs.gov.au/ausstats/abs@.nsf/94713ad445ff1425ca25682000192af2/1647509ef7e25faaca2568a90 0154b63?OpenDocument>

Baird D (1965) A fifth freedom? *British Medical Journal* **2**, 1141–1148.

Boyden S, Blanden R, Mims C (2013) Frank John Fenner FAA. 21 December 1914–22 November 2010. *Biographical Memoirs of Fellows of the Royal Society. Royal Society (Great Britain)* **59**, 125–144.

Chin G, Marathe T, Roberts L (2011) Population. *Science* **333**, 539–594.

Emmott S (2013) *Ten Billion*. Penguin Books, London.

20

Reflections on the Fenner Conference

Ian Lowe

Introduction and overview

It was fitting that the keynote address to the 2013 Fenner Conference was given by
Paul Ehrlich, as he was co-author of the landmark book which first drew together
these three critical issues (Ehrlich and Ehrlich 1970). As that book showed over
40 years ago, a growing human population compounded by increasing resource use
per person inevitably puts pressure on the natural environment. The obvious
conclusion of the contributions gathered in this book is that human civilisation
now faces a crisis. It is clearly impossible for all humans to use resources at the level
of contemporary Australians. It is equally impossible for Australia to continue on
the growth trajectory assumed by almost all politicians, even if we could ignore the
impacts of our consumption on the rest of the world. We are already seeing serious
environmental consequences of the current level of total consumption, but most
decision-makers assume and welcome both further population growth and
increasing per capita consumption, putting compounding pressure on natural
systems.

In opening the conference, Suzanne Cory called for planning the future with the aim of ensuring future prosperity by considering the population that could be sustainably supported. She also reminded us of the urgency of responding to climate change by 'substantial and sustained reduction of greenhouse gas emissions'. Ehrlich expanded on this theme, pointing out that 900 million people today are hungry and a further two billion are under-nourished, but climate change is likely to reduce overall food production. At the same time, increasing populations are causing a double squeeze on food production, with the need for living space reducing the availability of productive land just as the energy demands of the growing populations are accelerating climate change. Ending on a positive note, Ehrlich noted that societies can change very rapidly when the time is right, reminding his audience that the job of informed scientists is to 'righten the time': produce the 'knowings' that will motivate change.

The conference was a reminder that we have known about the emerging problems since the Club of Rome published its landmark report, *The Limits to Growth* (Meadows *et al.* 1972). It was the first attempt to model the long-term consequences of global trends, taking advantage of the increasing capacity of computers to process large quantities of data. The results were simplified by the media, which generally suggested that the report had concluded we were rapidly exhausting critical resources. What the report actually said was that if the existing trends of growth in population, resource use, industrial output, agricultural production and pollution were all to continue, we would reach limits within 100 years (that is, by 2070). Since it also concluded that these limits would lead to economic and social decline in the early to middle decades of the 21st century, the report should at least have provoked serious study and reflection, if not a revision of the standard development assumptions. Instead, it was attacked and belittled, usually by people who showed little sign of having actually read it, and for whom the very notion of limited growth was deeply offensive. The report concluded that it was entirely possible to shift the trajectory of human development onto a different path, one that would be sustainable into the distant future, but to do this would require new policy settings. That conclusion was also ignored.

In 1973, the world experienced the OPEC oil embargo and a consequent rapid increase in petroleum prices. While most observers had not anticipated this event, one expert had predicted it nearly 20 years earlier. M.K. Hubbert (1956) had calculated that US oil production would peak around 1971, leading to a need to import more oil and changing the balance between producers and purchasers in the world oil market. At the time, industry derided the very idea that oil resources were finite and that production could decline. Governments did not appear to notice the debate at all. Once the notion of peak oil had been verified at the national level, various analysts began using the same methodology to analyse global production. The calculation was not straightforward, as estimates of oil reserves were mostly proprietary information, kept confidential by either oil companies or national governments. While there was a good deal of uncertainty about the raw data for undertaking the analysis, by 1975 there was a broad consensus that world oil production would peak sometime between 2000 and 2020, with the most popular estimate being around 2010.

In 1985, the global scientific community warned that human activity was changing the global climate. The statement of the Villach Conference in that year related the observed changes in climate to the measured increases in greenhouse-gas concentrations; for the first time, climate scientists spoke up as a global body and suggested a relationship between human activity and the changing climate. The 1987 report of the World Council on Environment and Development, *Our Common Future*, considered the evidence of limited oil resources and the emerging consensus about climate change (WCED 1987). Recognising the fundamental importance of energy to modern civilisation, it concluded that new energy systems will be needed to power future human development, but noted that the changes would require 'new dimensions of political will and institutional cooperation'. In 1992, the Rio Earth Summit concluded the problem was sufficiently urgent to justify developing the Framework Convention on Climate Change. By 1997, the science had provided such convincing evidence of the problem that the global community agreed to the Kyoto Protocol. That agreement was hammered out despite concerted opposition from energy-intensive industries, the commercial world generally and a few rogue states like Saudi Arabia and Australia.

In the 1990s and early 2000s, a series of national and global reports spelled out the serious environmental problems we face. The first national report on the state of the environment (State of the Environment Advisory Council 1996) concluded that many aspects of the Australian environment were in good condition by international standards, but we had some very serious problems that would prevent us achieving our stated goal of developing sustainably: declining biodiversity, the state of much of our rural land, our inland rivers, pressures on the coastal zone and greenhouse-gas emissions. A second report, five years later, found that all of those serious problems had worsened. The third and fourth reports, at further five-year intervals, also concluded that the most serious problems were all getting worse (State of the Environment 2011 Committee 2011). There has been no concerted political response to these reports; if there is a trend, then it appears that the political emphasis on inappropriate forms of economic development has strengthened.

At the global level, the warning bells have been ringing for 20 years. Five reports in the United Nations Environment Programme's series Global Environmental Outlook (GEO), the report of the International Geosphere-Biosphere Programme, the Millennium Ecosystem Assessment research program and five reports by the Intergovernmental Panel on Climate Change have all warned that we are dangerously exceeding the capacity of natural systems. The second GEO report said explicitly, 'the present approach is not sustainable. Doing nothing is not an option.' The carefully argued conclusions of thousands of the world's best scientists – that we are stretching the capacity of natural systems to provide the ecosystem services on which life depends – had almost no impact on decision-makers. Even the urgent problem of climate change has only produced a half-hearted response. Australia's legislated Emissions Trading Scheme had inadequate targets and absurdly generous handouts to the worst polluters, but the newly elected Abbott Government in 2013 pledged to scrap the scheme because of its alleged cost burden on productive activity. So the timing of this book is critical: it constitutes an urgent wake-up call to the community and to decision-makers who are in denial about the scale of the problems we face and the urgency of responding.

Environmental implications of population growth

Hugh Possingham, David Lindenmayer and Chris Dickman provided three different but equally disturbing views of the consequences flowing from population growth. The evidence is clear: the dramatic growth in the human population in the last hundred years has been accompanied by an equally dramatic loss of biodiversity. The most extreme example is the loss of some 270 terrestrial vertebrate species, but wherever we look – birds, mammals, fish, reptiles – current rates of species loss are one to two orders of magnitude greater than the long-term average over the Earth's history. We are now in the sixth major extinction event, the only one driven by the demands of a single species rather than catastrophic perturbation of natural systems. Dickman argues that the urbanisation process causes 'cultural loss of memory' about our biological heritage and a 'diminishing connection with what is left', leading to the gloomy conclusion that many of our most charismatic species are likely to disappear this century. Thus there are some inevitable consequences of an Australia in which the human population continues to grow and the demands of that population continue to increase: an Australia with less biological diversity and diminished cultural richness. Lindenmayer pointed out that the failure to maintain biodiversity is a clear indication that we are not living in a way that is ecologically sustainable, although that has been Australia's declared goal for 20 years (COAG, 1992). Possingham showed that the Kuznets curve, widely assumed by economists to guarantee improving environmental quality with increasing wealth, does not work for biodiversity. The main policy problem with biodiversity is the long time lag between action, such as destroying habitat, and the inevitable consequences. If 90 per cent of habitat is destroyed, as it has been for significant communities in southern Australia, eventually 50 per cent of species will be lost, compared with the 5–10 per cent lost so far. On a positive note, Possingham argued that a trust fund created from a fortnight's defence budget ($16 billion) would provide enough income to save Australia's remaining biodiversity.

Social and economic implications of population growth

Bob Birrell, Mark O'Connor and Jane O'Sullivan showed that the current rate of population growth in Australia is having significant social and economic costs, although these are usually overlooked by those who claim there are offsetting benefits. Birrell noted that recent population growth has averaged 380 000 per year, with the net migrant intake varying between 172 000 and 316 000. He argued that the very high rates of migration are largely driven by employers putting pressure on governments, also noting a trend for migrants on temporary visas to be allowed to obtain permanent residence. O'Connor noted the parallel pressure to allow those who come to Australia to study to remain here after graduating, attributing a variety of social and economic problems in Australia to the 'third-world growth rate'. He also argued that environmental issues such as climate change are almost necessarily exacerbated by population growth: 'It's hard to reduce your footprint if you keep adding more feet.' O'Sullivan documented the infrastructure cost of increasing population and attacked the argument that high migration levels are needed to avoid the problems of an ageing population. The countries with the most obvious levels of ageing, Japan and Germany, now acknowledge 'depopulation dividends' with no downturn at all in the proportion of people employed, whereas the nations with the lowest average age are among the poorest on Earth. O'Sullivan concluded that the keys to prosperity in the modern world are 'natural resources per capita, the durability of man-made assets and the minimisation of economic rent payments', all of which are enhanced by stabilising the population.

Mineral resources

The issue of mineral resources reveals some interesting tensions, with obvious links to the environmental consequences of production. Michael Lardelli argued that the concept of 'peak oil' refers to the peaking of the rate of production, as has already happened for conventional oil. While the fossil fuel industry concentrates on the continuing development of the overall resource base, the critical issue for industrial society is the rate at which those resources can be developed. With our domestic oil now significantly depleted, Australia has 'a growing vulnerability to oil shortage shocks'. Simon Michaux developed a more fundamental argument, noting that the energy and water used per unit of mineral production is steadily increasing as richer deposits are exhausted and mining moves on to poorer grades of ore. In the last decade, there has been a 40 per cent increase in resources used per unit of product. So peak oil, or environmental limitations on the use of energy, raises the possibility that overall energy production could peak as early as 2017. This analysis raises the possibility of a more general problem that could be called 'peak mining' or a limit on the possible scale of minerals production. If population continues to grow, the available resources per person will peak and then decline. In the absence of available large-scale energy systems to replace fossil fuels, Michaux argued, 'the decline in production rates could well be a permanent trend'. Sharyn Munro showed that the current scale of mining coal is causing very serious environmental problems in such areas as the Hunter Valley, which now resembles a lunar landscape when seen from above, but governments continue to support further expansion of coal and coal seam gas production. The scale of devastation is causing 'unprecedented social resistance … from conservative rural people'. Taking into account resource limits, energy demands and growing resistance to social and environmental disruption, the overall prognosis for the minerals industry is not bright.

Climate change

Ian Dunlop developed an extension of this argument, noting that the 2005 peaking of conventional oil production 'has led to massive investment in unconventional alternatives such as tar sands, shale oil, shale gas, coal seam methane', all of which require more energy – and produce greater environmental impacts – per unit of delivered energy. As Michael Raupach argued, we are already facing the risk of dangerous climate change. While there has been a slowing in the rate of increase in average atmospheric temperature, there is no evidence that climate change is slowing, and no physical reason to expect it to slow while we continue to increase the atmospheric concentrations of the major greenhouse gases: carbon dioxide and methane. So the inevitable conclusion is that there is a limit to how much fossil fuels can be burned if we wish to avoid dangerous climate change. That means that we need 'to change the mix of energy resources away from fossil fuels, to limit population growth and wasteful resource consumption, and to keep a large proportion of fossil fuel reserves in the ground'. Dunlop expressed concern that the political influence of the fossil fuel industries is 'cutting off options to make the transition to a low-carbon society in good order'. To achieve an orderly transition, he argued, we need 'integrated climate and energy solutions'. In the absence of such integrated solutions, the attempts to resolve one serious problem – resource depletion – are certain to exacerbate another serious problem – climate change – by using more and more energy to produce smaller and smaller quantities of usable resources. Anthony McMichael showed that climate change 'threatens the ecological and social foundations of population health' by affecting food production, water supplies, disease patterns and other health risks. He concluded that a responsible approach involves both stabilising the population and curbing per capita consumption.

Food, land and water

Michael Jeffery argued that 'saving the soil' is the key to maintaining or increasing food production to meet the demands of a growing population. We need to recognise that soil, water and native vegetation are 'key national strategic assets, to be managed accordingly'. That in turn requires that farmers be recognised as 'primary carers of the agricultural landscape' and rewarded for that role. It also demands that urban Australia be re-connected with its rural roots, reversing the trend of recent decades. Rhondda Dickson argued that the significant decline of the Murray–Darling system (and its capacity to produce food) had been a consequence of the over-allocation of irrigation water and expressed confidence that the new Basin Plan would allow restoration of the system's natural values. Gary Jones drew lessons from the over-use of water in the Basin to discuss proposals for increasing agricultural production in northern Australia. Any such proposals need 'to integrate with the ebb and flow of seasonal rains and stream flows', so development will need to be selective and based on principles of 'broad environmental and cultural stewardship'. Rather than repeating the mistakes of the south, any new agricultural projects in northern Australia need to choose appropriate catchments and soils, then manage catchment and soil water balances to minimise ecological impacts. Provided these principles are followed, it will be possible to achieve modest expansion of food production in northern Australia.

Obstacles and opportunities

Stabilising population, reducing resource use and limiting environmental impacts are all issues that require social choices. So the final panel at the Fenner Conference addressed the social factors that will determine whether we meet the challenges we now face. Paul Collins noted the growing awareness of environmental problems in faith communities, but argued that 'Christians particularly are stymied by their anthropocentrism' and so mostly fail to take seriously the population challenge. He called for 'a new ethics' to replace the 'secular myths' of progress and 'perfect limitless markets'. Haydn Washington argued denial is 'a major obstacle to solving the entwined crises' of population growth, resource depletion and climate change. While denial of an uncomfortable reality is delusional, 'many people, cherish their delusions', especially the myth that growth can continue forever in a closed system. Kelvin Thomson lamented the absence of public debate about these critical issues, especially population, and gave notice of his intention to form a new NGO to focus on important long-term issues; Victoria First has since been launched. Julian Cribb offered a positive vision of 'social media taking over from traditional forms of controlled media', leading to the possibility of higher levels of understanding emerging through global virtual communities and catalysing a concerted community-led response. By contrast, he argued that governments are 'dinosaurs bogged in a tar-pit' of old attitudes and wilful ignorance, raising the alternative of a dystopia caused by failure to address the challenges.

Final declaration

Those present at the final session of the Fenner Conference drafted a declaration. While the process of drafting by a group of more than a hundred people inevitably sets limits on what can be agreed by consensus, a strong statement emerged:

Recognising

- the inextricable links between population, resources and climate change
- that human economic systems are dependent on natural ecological systems
- that a sustainable future depends on widespread ecological literacy
- that we are fast approaching the limits to resource (particularly oil) availability and scale of use
- that we face the prospect of catastrophic climate change that will affect society and the economy irrevocably
- that the world must keep four-fifths of fossil fuels in the ground, and
- that global population continues to grow by 80 million, and Australia's by 400 000 annually of which 60 per cent comes from net overseas migration ...

we therefore call on Australians and their governments to

- develop policies to stabilise Australia's population
- end the destruction of habitat
- develop plans to maintain domestic power, food production and distribution systems, and water and sanitation systems as fossil fuels peak and decline
- decarbonise our energy supply as a matter of urgency and develop clean energy renewable systems
- stop coal exports and end subsidies for exploration of oil, gas and coal
- incorporate the principle of ecological sustainability into core curricula at all levels
- make contraceptives more freely available and significantly increase the family planning component within Australia's foreign aid budget.

This seven-point plan would go a long way toward addressing the intertwined challenges. Within Australia, it would stabilise the population, recognise resource limits and take positive action to slow climate change, as well as educating the community to assist people to make responsible decisions. It would also increase our contribution to helping poorer countries in our region curb their population growth.

References

Council of Australian Governments (1992) *National Strategy for Ecologically Sustainable Development*. Department of the Environment, Council of Australian Governments, Canberra.

Ehrlich PR, Ehrlich AR (1970) *Population, Resources, Environment: Issues in Human Ecology*. W.H. Freeman, San Francisco.

Hubbert MK (1956) Nuclear energy and the fossil fuels. *Drilling and Production Practice* (American Petroleum Institute, 1956), 7–25.

Meadows DH, Meadows DL, Randers J, Behrens WW (1972) *The Limits to Growth*. Universe Books, New York.

State of the Environment 2011 Committee (2011) *Australia State of the Environment 2011 In Brief*. DSEWPAC, Canberra.

State of the Environment Advisory Council (1996) *State of the Environment Australia 1996*. CSIRO Publishing, Collingwood.

World Commission on Environment and Development (1987) *Our Common Future*. Oxford University Press, Oxford.

Index